Location Theory

Continued on inside back cover

Location Theory

Jean Jaskold Gabszewicz and Jacques-François Thisse
Université Catholique de Louvain, Belgium

Masahisa Fujita
University of Pennsylvania, USA

and

Urs Schweizer
University of Bonn, FRG

A volume in the Regional and Urban Economics section
edited by
Richard Arnott
Queen's University, Canada

harwood academic publishers
chur · london · paris · new york

Harwood Academic Publishers

P.O. Box 197
London WC2E 9PX
England

58, rue Lhomond
75005 Paris
France

P.O. Box 786
Cooper Station
New York, NY 10276
United States of America

Library of Congress Cataloging-in-Publication Data

Location theory.
(Fundamentals of pure and applied economics, ISSN 0191-1708; v. 5. Regional and urban economics section)
Includes bibliographies and index.
1. Theory-Location. 2. Land use, Urban. 3. Space in economics. I. Gabszewicz, Jean Jaskold. II. Series: Fundamentals of pure and applied economics; v. 5. III. Series: Fundamentals of pure and applied economics; v. 5. Regional and urban economics section.
HD58.L63 1986 338.6′042 86-12150
ISBN 3-7186-0297-0

ISBN 3-7186-0297-0. ISSN 0191-1708. Printed in the United Kingdom.

Contents

Introduction to the Series

Drawing on a personal network, an economist can still relatively easily stay well informed in the narrow field in which he works, but to keep up with the development of economics as a whole is a much more formidable challenge. Economists are confronted with difficulties associated with the rapid development of their discipline. There is a risk of "balkanisation" in economics, which may not be favorable to its development.

Fundamentals of Pure and Applied Economics has been created to meet this problem. The discipline of economics has been subdivided into sections (listed inside). These sections include short books, each surveying the state of the art in a given area.

Each book starts with the basic elements and goes as far as the most advanced results. Each should be useful to professors needing material for lectures, to graduate students looking for a global view of a particular subject, to professional economists wishing to keep up with the development of their science, and to researchers seeking convenient information on questions that incidentally appear in their work.

Each book is thus a presentation of the state of the art in a particular field rather than a step-by-step analysis of the development of the literature. Each is a high-level presentation but accessible to anyone with a solid background in economics, whether engaged in business, government, international organizations, teaching, or research in related fields.

Three aspects of *Fundamentals of Pure and Applied Economics* should be emphasized:

—First, the project covers the whole field of economics, not only theoretical or mathematical economics.

—Second, the project is open-ended and the number of books is not predetermined. If new interesting areas appear, they will generate additional books.
—Last, all the books making up each section will later be grouped to constitute one or several volumes of an Encyclopedia of Economics.

The editors of the sections are outstanding economists who have selected as authors for the series some of the finest specialists in the world.

J. Lesourne *H. Sonnenschein*

Spatial Competition and the Location of Firms

JEAN JASKOLD GABSZEWICZ and JACQUES-FRANÇOIS THISSE†

Université Catholique de Louvain, Louvain la Neuve, Belgium

On ne fait point de l'industrie entre ciel et terre; il faut se poser quelque part sur le sol.
L. Walras, *Eléments d'économie politique pure.*

1. INTRODUCTION

A competitive market is often viewed as a central exchange place where a large number of buyers and sellers are in free intercourse with one another. Arbitrage among sellers and buyers then leads to the competitive price. Examples are the Stock Exchange, the Chicago Wheat Market, and the periodic markets where traders meet and transfer commodities on the spot. Let us call such a market a "concentrated" market.

In the real world, however, many market activities are performed at dispersed points in space, even though these activities concern the same product. The market now operates over a complex network of scattered buyers and sellers, due to the topography of space and the location decisions of firms and households. Examples

† We are indebted to B. Allen, R. Arnott, M. Berliant, M. Fujita and L. Phlips for their helpful comments on a first draft of this article. Special thanks are due to S. Anderson for numerous suggestions and for careful editing which led to significant improvements to the paper. The financial support of the Fonds National de la Recherche Scientifique is also gratefully acknowledged. Part of this research was done while the second author was Visiting Professor at the University of Copenhagen.

of such "dispersed" markets abound in real life: think of department stores located in a town, gasoline stations located along a road, cement factories located in a country, etc.

Once the dispersed nature of many markets is recognized, then the natural companion assumption of a concentrated market, namely that it embodies numerous sellers and buyers, also needs revision. For a dispersed market it implies that each firm finds only a few rivals in its immediate neighborhood; further away there are more competitors but their influence is lessened by transportation costs. Similarly, not all consumers are alike to the firm; those who are far away will not buy from this firm because they have to pay too high a transportation cost. Thus "the market is commonly subdivided into regions within each of which one seller is in a quasi-monopolistic position" (Hotelling [50], p. 41). Under these circumstances, the possibilities of arbitrage are substantially reduced and firms enjoy some monopoly power. Consequently, in a dispersed market, the competitive assumption becomes hardly tenable and, as repeatedly advocated from Sraffa [91] to Greenhut [40], alternative approaches must be explored to describe the conscious interactions amongst few separated sellers and many scattered buyers.

The difference between the concepts of concentrated and dispersed markets has an analog, in industrial economics, in the difference between a market for a homogeneous product and an industry with differentiated products. In a market with a homogeneous product, substitutes are bunched into a single point of the space of characteristics, and sellers of this product may be numerous. In an industry with differentiated products, substitutes are dispersed in the space of characteristics, and the seller of a particular variety enjoys a quasi-monopolistic position relative to the buyers who most preferred it. Thus the interest in modelling spatial competition which arises from dispersed markets immediately extends to the process of competition amongst differentiated commodities.

The standard models of microeconomics do not seem well-suited to analyze dispersed markets (or industry with differentiated products). The partial equilibrium model is designed to characterize market solutions when the product is homogeneous; the general equilibrium model is designed to explain price determination, taking into account the interdependencies among all commodities. Between these two extremes lies an intermediate degree of complexity, represented by the model of spatial competition. Unlike partial equilibrium analysis, this model allows for substitution; unlike general equilibrium analysis, it deals only with the range of products which are taken as belonging to the same industry.

In a pathbreaking paper, Hotelling [50] provided the framework for the basic model of spatial competition. The present article rests upon this work, and aims to provide an integrated overview of the contributions rooted in the same tradition. It is organized as follows. After reviewing the main results obtained in the theory of spatial monopoly (Section 2), we propose a unified framework describing the basic ingredients of the process of competition in a dispersed market (Section 3). Then we investigate the market solution for a given number of firms (Section 4) and in the case of entry of new firms (Section 5). Finally, we briefly discuss some possible reformulations of the basic model and present some concluding remarks (Section 6).

2. SPATIAL MONOPOLY

In this section, we deal with the selection of a price-location pattern by a monopolist facing a finite system of demand functions dispersed over space. Two types of spatial price policy are considered, namely delivered pricing (2.1) and mill pricing (2.2). The former refers to a situation where the product is shipped to the consumers at the expense of the firm; in contrast, the latter supposes that customers move to the firm and bear transport costs.

2.1. Delivered pricing

Delivered pricing may take two forms. The firm may either charge a uniform price in all local markets irrespective of their location, or else set a specific price for each local market on the grounds of its location. Since transportation is under the control of the firm, consumers located at different places do not have to meet at the same market-place. Accordingly, it must be expected that—unless there are legal constraints—delivered pricing is often associated with some degree of discriminatory pricing.

The most extreme case arises when local markets are totally separated and no arbitrage can occur among customers at different places. Hence, assuming constant marginal production costs, maximum total profits are achieved when the spatial monopolist maximizes his profits on each local market separately.

Let $D_1(p_1), \ldots, D_m(p_m)$ be the demand functions for the monopolist's product in markets $s^1, \ldots, s^m$ in space S, respectively. Let $t(s, s^i)$ be the cost in terms of a given numéraire of transporting a unit of the product from the firm at $s \in S$ to market i. Assume that production entails a fixed set-up cost, f, and constant marginal cost, v, so that costs are given by $f + vq$.[1] Then total profits are written as

$$\sum_{i=1}^{m} p_i D_i(p_i) - v \sum_{i=1}^{m} D_i(p_i) - f - \sum_{i=1}^{m} t(s, s^i) D_i(p_i).$$

Clearly, for any given price vector $(\bar{p}_1, \ldots, \bar{p}_m)$, it is always profitable for the monopolist to choose a location s in S which minimizes its total transportation costs given by $\sum_{i=1}^{m} t(s, s^i) D_i(\bar{p}_i)$. In particular, if $(p_1^*, \ldots, p_m^*, s^*)$ is a profit-maximizing solution, then the location s^* must minimize the transport costs corresponding to the quantities $D_1(p_1^*), \ldots, D_m(p_m^*)$. To characterize the location chosen by the spatially discriminating monopolist, it is then sufficient to study the properties of s^* as a minimizer of $\sum_{i=1}^{m} t(s, s^i) D_i(p_i^*)$.

Following a well-established tradition in firm location theory, we now suppose that $t(s, s^i)$ is a linear function of the distance $\delta(s, s^i)$ between s and s^i, that is $t(s, s^i) = c\delta(s, s^i)$ with $c > 0$. In consequence, the profit-maximizing location s^* minimizes the function

$$T(s) = \sum_{i=1}^{m} c D_i(p_i^*) \delta(s, s^i).$$

This optimization problem—called the Weber problem[2]—has attracted the attention of many scholars in Operations Research. Not

[1] For simplicity of exposition, the production cost function is assumed to be independent of the firm location. However, the results of this section can be extended to cope with the transportation of inputs and the choice of an input mix.

[2] See Weber [98].

surprisingly, the properties of its solution depend on the type of distance considered.[3] Among the numerous mathematical models of distance used in location theory, we select the network distance.[4] Formally, a *network* is a subset of R^2 defined by the union of a finite number of arcs with a well-defined length (the arcs are "rectifiable") with intersect at most at their extremities; furthermore, the network is assumed to be of a single piece (the "connectedness" assumption). By definition, the set of *vertices* V of the network consists of the extremities of the arcs defining the network. Without loss of generality, we may assume that $s^i \in V$, $i = 1, \ldots, m$, and that any vertex which is not a market is a *node* of the transportation network, i.e., a point which is the extremity of at least three arcs. Finally, the distance $\delta(s, s')$ between any two points s and s' of the network is given by the length of the shortest route linking s and s'; δ satisfies the symmetry condition and the triangle inequality.[5] It should be clear that the characteristics of configuration, which are essential to real transportation spaces, are explicitly taken into account in the network model, thus showing its relevance for location theory.

A major property of the solution to the Weber problem with a network is then as follows: *the location* s^* (minimizing $T(s)$) *belongs to the set of vertices* V *of the network.*[6] This result, established independently by Guelicher [42] and Hakimi [43] and known as the Hakimi theorem, has several interesting implications. Firstly, it

[3] Unlike the O.R. models, most of the economic theory of firm location has been developed assuming the Euclidean (or crow-fly) distance. It is worthwhile noting that the theoretical results obtained with this distance are of little help for explaining the actual locations of firms. The essential reason is that the Euclidean distance gives rise to substitution effects which take place "in the small" whereas, in the real world, we observe substitutions "in the large;" see, e.g. Hoover [49] and Isard [53].

[4] Besides the Euclidean distance, the other models include the rectilinear distance (Huriot and Perreur [51]), the *lp*-distances (Love and Morris [65]), the central distances (Perreur and Thisse [75]) and the block norm distances (Ward and Wendell [97]). There are many "localization theorems" for the corresponding Weber problems; see, e.g. Durier and Michelot [22], Kuhn [61], Thisse, Ward and Wendell [95], Wendell and Hurter [99] and Witzgall [100].

[5] See Dearing and Francis [16] for a more detailed definition.

[6] Strictly speaking, s^* may be an interior point of an arc of the network when (and only when) the quantities shipped from the firm on both sides of s^* are equal. In such (zero-measure) cases, $T(s)$ is constant over the entire arc and is, therefore, minimized at each of its extremities.

shows that under spatial discriminatory pricing, the location problem of the spatial monopolist is of a *finite* nature, i.e., the set of points one should consider to get a profit-maximizing location can always be reduced to a finite subset of the network. Such a characterization has a great deal of intuitive appeal inasfar as locational decisions in (multi-) regional spaces are often taken as "discontinuous" (see Isard [53]). The solution to the monopolist's problem is then given by computing optimal prices for each local market for each location in V and choosing the point for which the corresponding profits are greatest. Secondly, we see that the profit-maximizing location is a *corner solution* in the sense that it is either a market or a node of the transportation network. Empirical observation supports this result (see Hoover [49]).[7–8]

The characterization provided by the Hakimi theorem can be made more precise when there is a "dominant" market, i.e., a market $j \in \{1, \ldots, m\}$ such that $D_j(p_j^*) \geqslant \sum_{i=1, i \neq j}^{m} D_i(p_i^*)$. In this case, it can be shown that s^j minimizes transportation costs and is thus the profit-maximizing location (see Witzgall [100]). This result therefore provides an explanation of the observed fact that some firms are established at a major market (see Hoover [49]).[9]

Until now, we have considered a single-plant firm. The following discussion addresses the possibility of the monopolist operating several spatially separated plants. Clearly, the existence of set-up costs places an upper bound on the number of plants; on the other

---

[7] The Hakimi theorem has been extended in a variety of directions; see Handler and Mirchandani [44] for a recent survey. Presumably, the most interesting one is that in which function t is only assumed to be increasing and concave in distance. Indeed, actual transfer costs of commodities generally increase less rapidly than in proportion to distance; see, e.g. Hoover [49] and Isard [53].

[8] In the case of a one-dimensional space, the Hakimi theorem implies that the profit-maximizing location is a market; see Norman [70]. Furthermore, when the transfer costs of the inputs are taken into account, the set of vertices to be considered consists of the markets, the input sources and the nodes; see Hanjoul and Thisse [45]. In the one-dimensional model, it then follows that the profit-maximizing location is a market or an input source; see Sakashita [79]. It is also worth pointing out that, in the case of colinear points, the monopolist's location is the median of the distribution of the output (and input) quantities multiplied by the corresponding transport rate; see, e.g. Witzgall [100].

[9] Other properties of the monopolist's location can be found in Hanjoul and Thisse [45].

hand, the larger the number of plants, the lower the transportation costs borne by the firm.

Since there are now several plants, the solution to the spatial monopoly problem should not only give the profit-maximizing price-location pattern, but should also specify the way in which production is allocated across plants. In other words, we want to determine the price policy p_i^*, $i = 1, \ldots, m$, the number n^* and the location s_j^*, $j = 1, \ldots, n^*$, of the plants as well as the fraction x_{ij}^* of the demand at i served by plant j at s_j^*, so as to maximize profits given by

$$\sum_{i=1}^{m} p_i D_i(p_i) - v \sum_{i=1}^{m} D_i(p_i) - nf - \sum_{i=1}^{m} \sum_{j=1}^{n} t(s_j, s^i) D_i(p_i) x_{ij}.$$

Interestingly, the approach used above can still be applied. Indeed, given $(p_1^*, \ldots, p_m^*)$, the profit-maximizing location-production pattern, (n^*, s_j^*, x_{ij}^*), must minimize the cost of production plus transportation,

$$nf + \sum_{i=1}^{m} \sum_{j=1}^{n} [t(s_j, s^i) + v] D_i(p_i^*) x_{ij}.$$

This problem has come to be known in Operations Research as the "simple plant location problem".[10] Thus, given the quantities of output sold on each market, the spatially discriminating monopolist chooses the socially optimal location-production pattern.[11] This is easy to understand since, under delivered pricing, the firm is able to internalize the whole (social) benefit generated by the choice of the most efficient pattern. Of course, this does not mean that the monopolist's plant location pattern would be identical to the one selected by a welfare maximizer. It is well known, indeed, that the quantities sold by the profit-maximizing monopolist are not socially optimal.

Does the Hakimi theorem hold for the multiplant firm? Clearly, for any given n^* and x_{ij}^*, location s_j^* must minimize the transport cost function $\sum_{i=1}^{m} t(s_j, s^i) D_i(p_i^*) x_{ij}^*$ relative to the jth plant so that, under the assumption of a constant transportation rate c, the

[10] See Manne [67]. For recent developments, see Cornuejols, Nemhauser and Wolsey [11], Erlenkotter [32] and Krarup and Pruzan [60].

[11] Heal [47] reaches a similar conclusion in a related model.

Hakimi theorem can be applied to each plant separately. Hence, the profit-maximizing locations are again contained in the set of markets and nodes.

To conclude, let us note that the above developments remain valid when price discrimination is limited by arbitrage and/or legal constraints. Indeed, given the profit-maximizing price(s), the firm still locates so as to minimize costs. In particular, this implies that, under uniform delivered pricing, the spatial monopolist sets up his plants at "corner points" of the transportation network.

2.2. Mill pricing

When customers bear transportation costs, discriminatory practices are not feasible as the firm location is now the market-place. (Assuming that the firm exercises discriminatory prices, consumers could proceed to the arbitrages at the firm's door without incurring additional transportation costs.) In what follows, we therefore assume that the monopolist charges the same mill price to its customers irrespective of their location.[12]

If the firm operates a single outlet, its profits can be written as

$$(p - v) \sum_{i=1}^{m} D_i(p + t(s^i, s)) - f.$$

We immediately notice that, in general, for a given price $\bar{p}$ the monopolist does not choose a location s in S which minimizes (consumers') transportation costs. Instead, he locates at a point that maximizes his total sales. The reason is that the location of the firm directly influences consumer demand. As a result, the argument developed in 2.1 ceases to be valid. This suggests that, in turn, the

[12] What are the reasons which lead a monopolist to choose mill pricing? There are several of them, from which we pick the following ones. Firstly, price discrimination, and spatial price discrimination is a special case of it, is illegal in several countries (including U.S. and E.E.C. countries). This implies that the spatial monopolist must choose between mill pricing and uniform delivered pricing (or a combination of them). The relative superiority of these two price policies for a profit-maximizing firm does depend on the demand functions; see Stevens and Rydell [93]. In some cases, it is therefore optimal for the monopolist to choose mill pricing. Secondly, even when legal constraints are not binding, the profit-maximizing firm may elect not to price-discriminate when customers have access to a more efficient transport technology than the firm; see Gronberg and Meyer [41].

Hakimi theorem would no longer be valid in the case of mill pricing. Actually, it is still true that, at the profit-maximizing solution, the firm locates either at a demand place or at a node of the network when demand functions are decreasing and *convex.*[13] However, when the convexity assumption is relaxed, the set of vertices V may not contain a profit-maximizing location. To illustrate this, consider the following example. There are two consumers located respectively at the extremities of the line segment $[0, 1]$; the demand function of consumer i is given by max $\{0, \alpha - (p + t(s^i, s))^2\}$ where α is a positive constant; finally the transportation rate is equal to 1 and there is no production cost. Assuming that the firm is at $\theta \in [0, 1]$, the profit function is $p\{\alpha - (p + \theta)^2 + \alpha - [p + (1 - \theta)]^2\}$ for $p < \sqrt{\alpha} - 1$. It is then easy to see that, for a sufficiently large α, the unique profit-maximizing location is at $\theta = \frac{1}{2}$.

When the above model is extended to the case of a firm operating several outlets, profits become

$$(p - v) \sum_{i=1}^{m} D_i\left(p + \min_{j=1,\ldots,n} t(s^i, s_j)\right) - nf$$

(consumers are assumed to patronize the outlet with the lowest full price; for more details, see 3.2).

It is worthwhile pointing out that, at the profit-maximizing solution, the monopolist *does not* produce the corresponding quantities

$$D_i\left(p^* + \min_{j=1,\ldots,n^*} t(s^i, s_j^*)\right)$$

at the minimum production cost. Indeed, a single outlet is obviously the production cost-minimizing solution for the firm (recall that transport costs are borne by the customers). However, the monopolist is often induced to establish more than a single outlet, simply because this reduces transport expenses of the consumers which, in turn, leads to a higher demand and, thus, higher revenues for the firm. In addition, total production *and* transportation costs are not minimized, given the profit-maximizing quantities sold to the customers at $s^1, \ldots, s^m$. This is so because, unlike the delivered

[13] See Hanjoul and Thisse [45].

price case, the spatial monopolist is here unable to fully appropriate the benefits that accrue at a cost-minimizing configuration.[14]

Interestingly, the above discussion casts some doubt on the general claim that discriminatory pricing should always lead to a larger "deadweight loss" than mill pricing. Indeed, the additional cost generated by the excess capacity required to attract customers, could well exceed the welfare gains related to mill pricing.[15]

A few general conclusions emerge from the above analysis. First, the spatial price policy exerts an influence on the monopolist's locational decision. Results obtained under (uniform) delivered pricing and mill pricing are very different.[16] With delivered prices, one expects the location to be a market or a node of the network. In contrast, with mill prices, some "intermediate" locations are possible. Second, the choice of a particular price policy affects the efficiency of the monopolist's production-location pattern. Given the quantities sold to the customers at $s^1, \ldots, s^m$, total transportation and production costs are minimized under delivered pricing, but not under mill pricing.

3. THE BASIC INGREDIENTS OF SPATIAL COMPETITION

We now move to spatial competition stricto sensu, to the extent that we explicitly consider several firms interacting within the same industry. In this section, we discuss successively the concept of industry (3.1), the structure of demand (3.2) and the definition of equilibrium (3.3).

3.1. The concept of industry

The spatial framework is particularly suited to defining a concept of industry relying on the geographical distribution of firms and consumers. Consider a finite set $N = \{1, \ldots, n\}$ of firms producing a given homogeneous product, and a finite set $M = \{1, \ldots, m\}$ of consumers. Each firm j, $j \in N$, is located at some point s_j in space S;

[14] See Heal [47] and Kats [57] for related arguments.
[15] See Phlips [76] for a complementary discussion of this topic.
[16] This confirms Greenhut [39] 's thesis. See also Section 6.

similarly, each consumer i, $i \in M$, is located at $s^i \in S$. Denote by $t(s^i, s_j)$ the cost of transporting a unit of the good between consumer i's and firm j's locations, measured in terms of a given numéraire. Assume that the product is produced by firm j at constant marginal cost c_j, also measured in terms of the numéraire. Finally let us suppose throughout this section that each consumer either does not buy the product, or buys exactly a single unit of it (per unit of time). Thus, assuming that the number of firms is finite, consumer choice operates within a finite set of mutually exclusive alternatives. Define π_i as the highest price that consumer i is willing to pay to obtain a unit of the considered commodity, i.e., π_i is the reservation price of consumer i.

Not all consumers are potential customers of a given firm. Indeed, it may happen that some are located so far away from it that their reservation price cannot cover the sum of the transportation cost and marginal production cost. Accordingly, even at a zero profit price, such consumers will never buy from such firms. This observation motivates the following definition. The *potential market* of firm j, denoted by M_j, is the set of all consumers $i \in M$ for which $c_j + t(s^i, s_j) \leqslant \pi_i$. The notion of potential market suggests an intuitive idea of potential competition between firms. In particular, if, for two firms, the corresponding potential markets intersect, we say that these two firms are *potential direct competitors*. Indeed, when two firms are potential direct competitors, there always exist prices which exceed or equal marginal costs such that some consumers consider buying the product from one or the other of these firms at these prices (obviously, these consumers must belong simultaneously to the potential markets of both firms).

The notion of potential direct competition is not sufficient to capture the whole structure of competition among firms, in particular between firms which are *not* potential direct competitors: we must also take into account the possibility of a potential indirect competition between two such firms. More precisely, we say that two firms are *potential indirect competitors* if (i) they are not potential direct competitors, and (ii) there exists a chain of firms in N linking them, and such that two subsequent firms in the chain are potential direct competitors. When two firms are potential indirect competitors, whatever the pair of prices they may choose, no consumer considers buying the product from one or the other of these firms. Nevertheless, each one of them, through a change in its own price, may still influence the market share of the other, through the price reactions of the intermediate firms in the chain.

In conclusion, we define an *industry* as any subset I of firms in N such that any pair of firms in I are potential direct or indirect competitors, while no firm in I is a potential direct or indirect competitor of a firm in $N-I$. The process of competition which underlies this definition can be visualized as a graph whose vertices correspond to locations of firms in N, and edges to pairs of firms which are potential direct competitors. Within this framework, an industry is the set of firms belonging to the same connected component of the graph. Adjacent vertices correspond to potential direct competitors, while vertices linked through some intermediate vertices correspond to potential indirect competitors. Particular specifications of this graph reflect different varieties of potential competition.

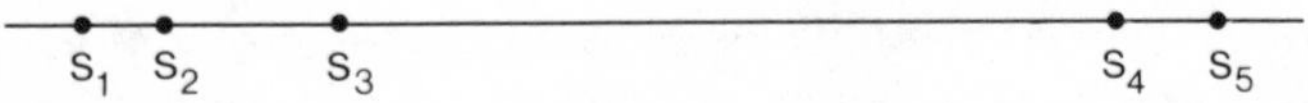

FIGURE 1.

As a first example, consider a geographical configuration with five population centers located along a road as in Figure 1, with a single bakery in each population center. Assuming that all consumers have the same reservation price π for a bread, and that the marginal cost c_j is the same for all bakeries, and equal to c, we see that, if $c+t(s_3, s_4)>\pi$ and $c+t(s_1, s_3)>\pi$, while $c+t(s_1, s_2)\leqslant\pi$, $c+t(s_2, s_3)\leqslant\pi$ and $c+t(s_4, s_5)\leqslant\pi$, bakeries $\{1,2,3\}$ define an industry, whereas bakeries $\{4,5\}$ define another one. Moreover bakeries 1 and 2, and 2 and 3, are potential direct competitors, while bakeries 1 and 3 are potential indirect competitors in the first industry; bakeries 4 and 5 are potential direct competitors in the second industry. Accordingly the corresponding graph is as in Figure 2. Although all five bakeries might seem, at first sight, to belong to the same industry, they do not in our approach, since no pair of firms, with the first extracted from $\{1,2,3\}$ and the second from $\{4,5\}$, are potential, direct or indirect, competitors.

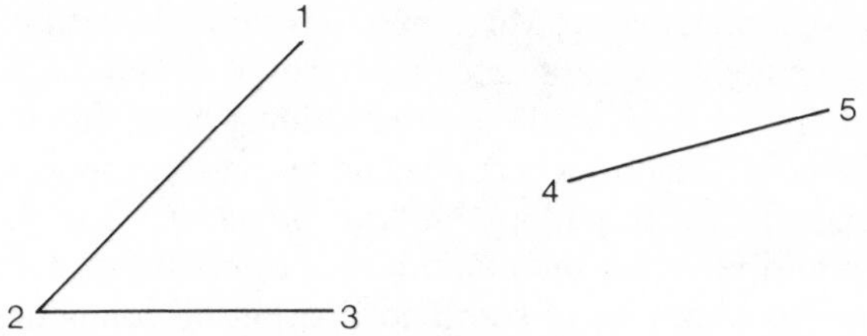

FIGURE 2.

A second example is provided by bakeries (and population centers) located on a star-shaped network as in Figure 3. If $c + t(s_j, s_h) > \pi$, $\forall j, h, j \neq 3, h \neq 3$, but $c + t(s_j, s_3) \leq \pi$ for all $j \neq 3$, we easily see that 3 and j are all potential direct competitors while j and h are potential indirect competitors. Accordingly, in this case, there is only one industry, and all firms are members of this industry.

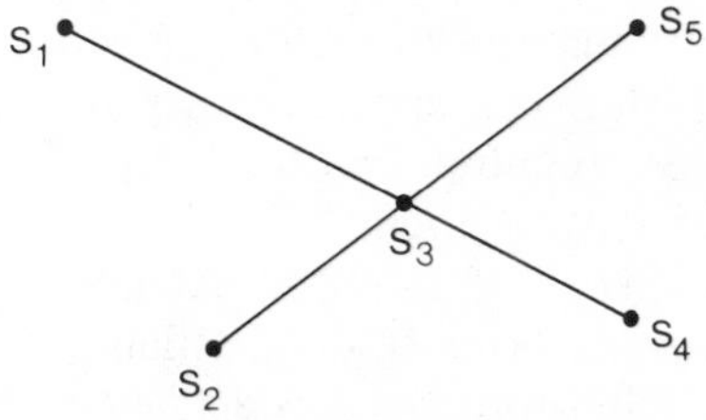

FIGURE 3.

Finally a still more complex structure of potential competition is provided by the example depicted in Figure 4, where it is assumed that $c + t(s_i, s_j) \leq \pi$ for any pair (i, j). In this case, any pair of firms are potential direct competitors.

At the root of the above concept of industry lie the consumers'

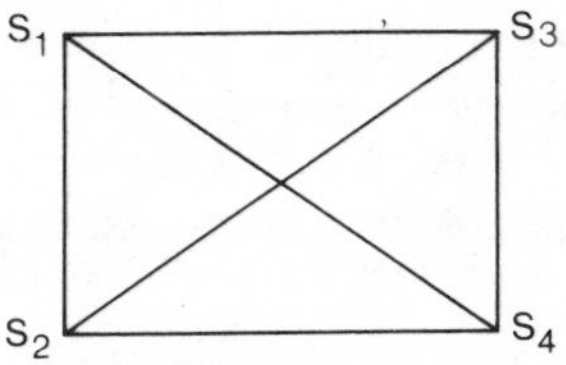

FIGURE 4.

and firms' characteristics. More precisely, our definition takes the following elements as parameters: (i) the geographical distribution of consumers, as well as their reservation prices; (ii) the number and location of firms; and (iii) production and transportation costs. Any change in these parameters can affect the structure of the industry and, therefore, the potential for competition. For example, the entry of a new firm at some appropriate location can link two preexisting and separated industries into a single one. This is illustrated with the aid of Figure 1 when a new firm, say firm 6, establishes itself at s_6 between firms 3 and 4 in a way that $c + t(s_3, s_6) \leq \pi$ and $c + t(s_6, s_4) \leq \pi$. Clearly, in such a case, the whole structure of potential competition is changed by the entry of this new firm and a stronger rivalry is expected.[17]

It is also interesting to compare our approach with respect to standard partial and general equilibrium models. With the former it shares the fact that all the goods which are not in the industry, are combined into a single composite commodity; with the latter it shares the property that the markets of all goods within the industry are interrelated. Thus our approach may be viewed as a kind of "intermediate" or "multiple partial" equilibrium model representing the intuitive usual notion of an industry.

Finally, our concept of industry extends in a straightforward manner to cope with product differentiation. Here the potential market of firm j, selling product j, is defined as the set of consumers i whose reservation price π_i^j for product j exceeds or equals the marginal cost c_j of firm j. Then the concept of industry is formulated as above using this alternative definition of a potential market. Interestingly enough, the spatial model reduces to a particular case of that model of product differentiation, with $\pi_i^j = \pi_i - t(s^i, s_j)$.[18]

3.2. The structure of demand in the industry

So far we have not yet considered the way in which industry price structure determines the demand assigned to each firm as a function

[17] Other examples of parametric changes could be similarly considered: decreases in transportation and/or production costs, due to technological progress, and increases in reservation prices, corresponding to changes in consumers' tastes and/or income, are also expected to enhance the potential of competition.

[18] See also Archibald, Eaton and Lipsey [2].

of prices. Assume from now on that firms follow a *mill price policy*. Given an industry consisting of n different firms, denote by $(p_1, \ldots, p_j, \ldots, p_n)$ the vector of mill prices in that industry. Then consumer i buys from firm j if and only if the following conditions hold:

i) $p_j + t(s^i, s_j) \leq \pi_i$;

ii) $p_j + t(s^i, s_j) = \text{Min}_{k=1}^{n}\{p_k + t(s^i, s_k)\}$;

iii) when $k \neq j$ exists such that $p_j + t(s^i, s_j) = p_k + t(s^i, s_k)$, then $t(s^i, s_j) < t(s^i, s_k)$ or $j < k$.

Condition (i) guarantees that consumer i benefits from buying from firm j; condition (ii) states that this benefit is maximal when the full price is minimized; finally, condition (iii) is a conventional device to assign consumers indifferent between purchasing from at least two firms, if any, to a single one of them. Therefore, on the basis of a comparison of prices and transportation costs, consumer i chooses to patronize firm j at the exclusion of any other firm in the industry. This corresponds to a rule of *mutually exclusive choices*.[19]

The above conditions express the basic principles which permit one to split the population of consumers into the customers of each firm and into the consumers who prefer not to operate in this particular market. More precisely, we define the market area of firm j at price $p_1, \ldots, p_j, \ldots, p_n$, denoted $A_j(p_1, \ldots, p_j, \ldots, p_n)$, as the set of consumers $i \in N$ for which conditions (i), (ii) and (iii) hold. Let $A_0(p_1, \ldots, p_n)$ denote the set of consumers not purchasing the industry good, i.e.,

$$A_0(p_1, \ldots, p_n) \underset{\text{def}}{=} \{i \in M \mid \forall j = 1, \ldots, n, p_j + t(s^i, s_j) > \pi_i\}.$$

Then the sets $A_j(p_1, \ldots, p_j, \ldots, p_n)$, $j = 0, \ldots, n$, form a partition of the set M of consumers. This partition is conditional upon the price vector $(p_1, \ldots, p_n)$. Any change in one or several prices can induce modifications in the market areas of firms in that some consumers shift from one firm to another, or choose to refrain from purchasing from any firm, or to buy from one of them.

Since each consumer buys at most a single unit of the product, it

[19] Interestingly, that rule still holds even if the consumers want to buy several, and not necessarily one, units of the product; see Novshek and Sonnenschein [73].

is natural to identify the demand addressed to firm j at prices $(p_1, \ldots, p_n)$, with the cardinal of the set $A_j(p_1, \ldots, p_j, \ldots, p_n)$. We denote by $D_j(p_1, \ldots, p_n)$ the corresponding number defining the *demand* to firm j at prices $(p_1, \ldots, p_n)$.

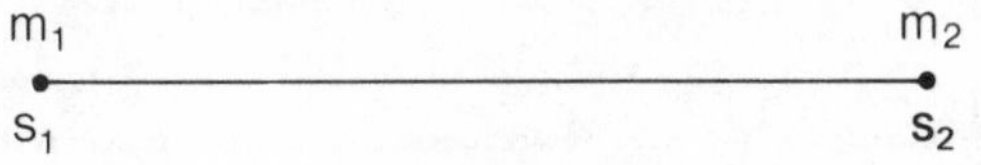

FIGURE 5.

Let us now illustrate the notion of demand by the following simple example. Assume that two groups of identical consumers with reservation price π, and two identical firms with marginal cost $c = 0$, are located on the road as in Figure 5.

Firms 1 and 2 are located, respectively, at s_1 and s_2, while m_1 consumers are located at s_1, and m_2 at s_2. Figure 6 provides a typical illustration of demand to firm 1 conditional on some price $\bar{p}_2$ fixed at a level such that $t(s_1, s_2) < \bar{p}_2 < \pi < \bar{p}_2 + t(s_1, s_2)$. There are three regions. In the first one, defined by $p_1 > \pi$, the firm has a zero demand. In the second region, corresponding to $\bar{p}_2 - t(s_1, s_2) \leqslant p_1 \leqslant \pi$, firm 1 supplies the consumers established at s_1. Finally, in the third region defined by $0 \leqslant p_1 < \bar{p}_2 - t(s_1, s_2)$, firm 1 attracts all consumers at s_1 and s_2, i.e., the consumers located in its local market as well as the customers of its rival.

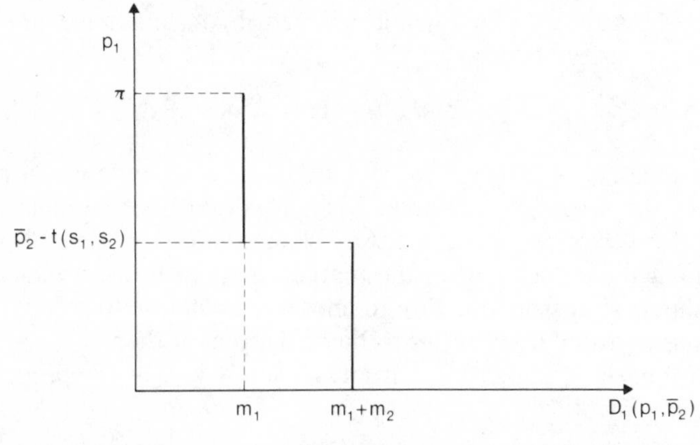

FIGURE 6.

Thus we observe in this example that the demand function to a firm may be highly *discontinuous*: over some ranges of prices, the demand is constant while there are some critical price values where a whole group of consumers shifts from one firm to another. Such discontinuities in the firms' demand functions are similar to those encountered in Bertrand [6]. They stem from the rule of mutually exclusive purchases by the consumers, combined with the atomic distribution of consumers over space. Indeed the former entails the discontinuity of the consumers' demands while the latter prevents the firm's demand from recovering continuity by aggregation. There lies the root of Hotelling's [50] brilliant idea to restore continuity in the firm demands by considering a nonatomic distribution of consumers.[20] More precisely, the demand side of the model is reformulated as follows. There is a continuum of consumers—all consumers located at $s \in S$ having the same reservation price $\pi(s)$—distributed over space according to a continuous density function $\Delta(s)$. The demand to firm j at prices $(p_1, \ldots, p_n)$ is then given by

$$D_j(p_1, \ldots, p_n) = \int_{A_j(p_1, \ldots, p_n)} \Delta(s)\, ds$$

where $A_j(p_1, \ldots, p_n)$, the market area of firm j, is now defined as the set of locations $s \in S$ for which

i) $p_j + t(s, s_j) \leq \pi(s)$;

ii) $p_j + t(s, s_j) = \mathrm{Min}_{k=1}^{n}\{p_k + t(s, s_k)\}$;

iii) when $k \neq j$ exists such that $p_j + t(s, s_j) = p_k + t(s, s_k)$, then $t(s, s_j) < t(s, s_k)$ or $j < k$.

To illustrate this representation, we consider the following example inspired from Hotelling [50]. It is assumed that identical consumers ($\pi = \pi(s)$ for all $s \in S$) are evenly spread out along the road linking s_1 and s_2 with a unit density, while the two firms are still located at s_1 and s_2. Furthermore, $t(s, s') = c\,|s - s'|$ for any two locations s and s' in $[s_1, s_2]$. The demand to every firm is then given by the (Lebesgue) measure of its market area at prices (p_1, p_2). For

[20] On this subject, see also Archibald, Eaton and Lipsey [2] and the discussion in 5.1.

instance, for $c(s_2-s_1)<\bar{p}_2<\pi<\bar{p}_2+c(s_2-s_1)$, the following regions of the demand to firm 1 may be distinguished

i) for $p_1>\pi$, $D_1=0$;

ii) for $2\pi-\bar{p}_2-c(s_2-s_1)\leqslant p_1\leqslant\pi$, $D_1=(\pi-p_1)/c$, i.e., these consumers for whom the surplus obtained from firm 2 is negative;

iii) for $\bar{p}_2-c(s_2-s_1)\leqslant p_1<2\pi-\bar{p}_2-c(s_2-s_1)$,

$$D_1=\frac{\bar{p}_2-p_1+c(s_2-s_1)}{2c},$$

which includes some consumers who would otherwise purchase from firm 2;

iv) for $p_1<\bar{p}_2-c(s_2-s_1)$, $D_1=s_2-s_1$, i.e., the whole market is served by firm 1.

This demand function is depicted in Figure 7. Clearly the demand to the firm is now continuous: as firm 1 reduces its price, consumers are gradually attracted by this firm, but no group of consumers suddenly shifts to it. However, the continuous representation is *not* sufficient to guarantee demand continuity. In particular, as will be seen in 4.1, the demand to a firm is still discontinuous when firms are located inside $[s_1, s_2]$.

The following remarks are in order. First, it should be apparent that the demand addressed to a firm does not necessarily depend on

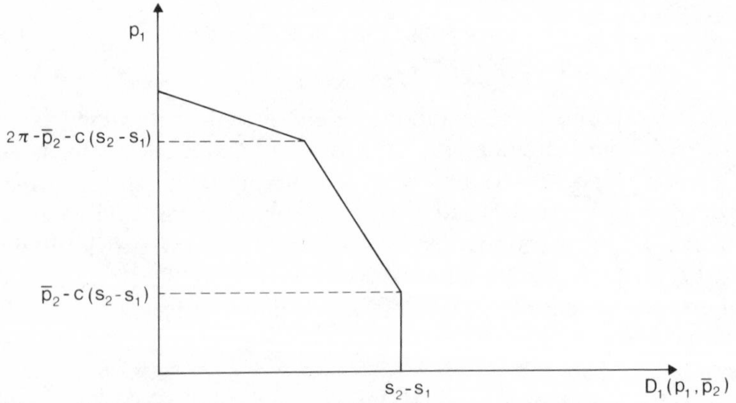

FIGURE 7.

the prices set by all firms in the industry. Indeed, given our definition of potential indirect competition, the demand to firm j is never a function of the prices charged by its potential indirect competitors. On the other hand, there always exist price levels for which D_j does depend on the prices chosen by any of its potential direct competitors. However, this dependence need not occur for all potential direct competitors simultaneously. It is determined by the relative position of firms in the transportation network. Thus, as pointed out by Kaldor [56], the cross-price elasticity between some pairs of firms belonging to the same industry is zero. Analogous to the concept of potential competitors introduced in 3.1, we say that, for a given price vector, two firms are *direct competitors* if their cross-price elasticity is nonzero; two firms are *indirect competitors* if (i) they are not direct competitors, and (ii) there exists a chain of firms of the same industry linking them, and such that each two subsequent firms in the chain are direct competitors.[21]

Second, looking at the demand functions of two firms individually may not be sufficient to determine whether those firms belong to the same industry. In most relevant cases, the presence of some "chain effect" linking together apparently independent firms seems to be inherent to the spatial (or product differentiation) framework. The whole demand system must then be inspected to delineate the industry. This is in sharp contrast to the standard interpretation and indicates how intricate an industry may be.

Third, and last, in the same manner that the structure of the industry is conditional upon the set of parameters which specify it, so does the structure of demand depend upon these parameters. In particular, it should be expected that the entry of a new firm into a given industry should lead to a complete reformulation of demands to existing firms after entry.

3.3. The definition of equilibrium in the industry

Given the above description of the industry and its demand structure, it remains to discuss the determination, via spatial competition, of prices and quantities in the industry. First, the very

[21] A similar idea has been used in a different context and for a different purpose, by Aumann [3].

nature of spatial competition hardly makes the perfectly competitive mechanism a realistic way to determine prices. Indeed a large number of firms in the industry is not sufficient to guarantee perfect competition; perfect competition also requires a large number of firms in each place where any one of them is located. In our framework one cannot avoid the difficult problem of conscious interaction among firms. In essence, space endows each firm with some degree of monopoly within its potential market. To the extent that the potential market of a firm intersects the potential markets of some other firms, its relative monopoly position is weakened by the countervailing possibilities open to these other firms. In what follows, we choose to concentrate on a *noncooperative* approach to price determination with independent firms.[22]

Assume that each firm in the industry knows how its demand depends on its own price and on those of its competitors. What could be an "equilibrium price system" in such an interacting decision context? Clearly any notion of equilibrium which would allow some firm, given the prices of the others, to increase its profit by changing its price could hardly be viewed as a satisfactory concept: in such a case there would exist an incentive for this firm to move away from its "equilibrium" position; but then, why would it be an equilibrium? Thus any satisfactory notion of equilibrium must avoid this drawback, and require, at the very least, the absence of such incentives at the equilibrium price system. In other words, a necessary property of equilibrium is that no firm can increase its profit by a unilateral change in price. This condition of internal consistency motivates the following notion of equilibrium: a *price equilibrium* is a n-tuple $(p_1^*, \ldots, p_j^*, \ldots, p_n^*)$ of prices such that $\forall j$, $j = 1, \ldots, n$, $\forall p_j \geqslant 0$,

$$P_j(p_1^*, \ldots, p_n^*) \underset{\text{def}}{=} (p_j^* - c_j) D_j(p_1^*, \ldots, p_j^*, \ldots, p_n^*)$$

$$\geqslant (p_j - c_j) D_j(p_1^*, \ldots, p_j, \ldots, p_n^*) \underset{\text{def}}{=} P_j(p_1^*, \ldots, p_j, \ldots, p_n^*).$$

[22] Of course the rivalry among firms could be solved by total or partial collusion. If collusion is total, then we are led back to the problem of spatial monopoly, considered in Section 2. Moreover, the case of partial collusion raises problems similar to those encountered in a noncooperative approach; for instance, Friedman [33] considers the noncooperative static equilibrium as the threat point of a dynamic collusive equilibrium. See also Palfrey [74].

Therefore a price equilibrium is by definition a Nash equilibrium of a noncooperative game whose players are firms, strategies are prices, and payoffs are profits. The price equilibrium not only reflects the rivalry among firms which are potential direct competitors, but is built on the whole net of both potential direct and indirect competition. To see this, consider the three firms 1, 2, and 3 depicted in Figure 1. Then, since 2 and 3 are potential direct competitors, p_3^* should depend, among other things, on the value of p_2^*; similarly, p_2^* is also dependent on p_1^*; as a result p_3^* depends on p_1^*, albeit 1 and 3 are potential indirect competitors only.

Let us illustrate the notion of price equilibrium with the two examples given in 3.2. If a price equilibrium exists in the case of the first example (see Figure 6), (p_1^*, p_2^*) say, it must be the case that both firms have a positive profit at (p_1^*, p_2^*) since they can always secure their local market by charging a positive price smaller than the transportation cost $t(s_1, s_2)$. Consequently, we must have $|p_1^* - p_2^*| \leqslant t(s_1, s_2)$. Assume, first, that $|p_1^* - p_2^*| = t(s_1, s_2)$. In this case, for $p_2^* = p_1^* + t(s_1, s_2)$ say, firm 1 can increase its profit by charging a price $\bar{p}_1 = p_1^* - \varepsilon$, with $\varepsilon > 0$ arbitrarily small; this contradicts the fact that p_1^* maximizes firm 1's profit against p_2^*, a property of price equilibrium. Suppose now that $|p_1^* - p_2^*| < t(s_1, s_2)$. If $p_1^* < \pi$, where π is the unique reservation price, then firm 1 can enjoy a higher profit by choosing any price $\bar{p}_1 = p_1^* + \varepsilon \leqslant \text{Min}\{\pi, p_2^* + t(s_1, s_2)\}$, again a contradiction. Furthermore $p_1^* > \pi$ implies $D_1 = 0$. As a result, p_1^* must be equal to π. A symmetric argument shows that $p_2^* = \pi$. Let us show that $p_1^* = \pi$ and $p_2^* = \pi$ is *not* a price equilibrium when

$$\pi > \frac{m_1 + m_2}{\text{Max}(m_1, m_2)} t(s_1, s_2).$$

If

$$\pi > \frac{m_1 + m_2}{m_2} t(s_1, s_2),$$

say, then firm 1 can increase its profit by undercutting firm 2's price. Indeed, for $p_2^* = \pi$, the corresponding profit of firm 1 is equal to $(\pi - t(s_1, s_2) - \varepsilon)(m_1 + m_2)$ which, for $\varepsilon > 0$ sufficiently small, is larger than the profit earned by that firm at $p_1^* = \pi$. Accordingly, in our first example, no price equilibrium exists when the reservation price π is large relative to the transportation cost.[23] On the other hand, when

$$\pi \leqslant \frac{m_1 + m_2}{\text{Max}(m_1, m_2)} t(s_1, s_2),$$

no firm has an incentive to undercut its competitor so that $p_1^* = \pi$ and $p_2^* = \pi$ is a price equilibrium. In other words, when the reservation price is small enough, both firms behave like local monopolists.

Consider now the second example given in 3.2 (see Figure 7). Given the shape of the demand function, it is easy to see that the profit function of firm i is concave in p_i over the domain $\{p_i; P_i(p_i, \bar{p}_j) > 0\}$ for all given $\bar{p}_j$. In consequence, the first-order conditions are also sufficient. Three cases may arise (contrast to the first example). In the first one, each firm maximizes its profit independently of the other, which means that the firms are concerned with the monopoly region of their demands. The first-order conditions are therefore given by

$$\frac{dP_1}{dp_1} = \frac{\pi - 2p_1^*}{c} = 0 \quad \text{and} \quad \frac{dP_2}{dp_2} = \frac{\pi - 2p_2^*}{c} = 0,$$

which yields $p_1^* = p_2^* = \pi/2$. This is a price equilibrium provided that the market areas at p_1^* and p_2^* are disjoint, i.e., $\pi < c(s_2 - s_1)$. (Compare this with the first example.) In the second case, each firm in choosing its profit-maximizing price compete with the other. This implies that the firms are now concerned with the competitive region of their demands. Hence the first-order conditions

$$\frac{dP_1}{dp_1} = \frac{-2p_1^* + p_2^* + c(s_2 - s_1)}{2c} = 0$$

and

$$\frac{dP_2}{dp_2} = \frac{p_1^* - 2p_2^* + c(s_2 - s_1)}{2c} = 0$$

must hold, i.e., $p_1^* = p_2^* = c(s_2 - s_1)$. In contrast to the first example, it never pays for one of the firms to undercut its competitor.

23 This is reminiscent of the price cycle obtained by Edgeworth [31] in the case of duopolistic competition with capacity constraints.

It remains to check that market areas are not disjoint at p_1^* and p_2^*, that is $\frac{3}{2}c(s_2-s_1)\leqslant\pi$. The third case deals with the remaining zone: $c(s_2-s_1)\leqslant\pi<\frac{3}{2}c(s_2-s_1)$. Elementary manipulations show that each firm maximizes its profit at the kink of its demand function so that market areas just touch. Accordingly, we have

$$p_1^*=p_2^*=\pi-\frac{c}{2}(s_2-s_1).$$

The preceding discussion leads to the following conclusions. First, when the reservation price is "low enough" compared to the transportation (and/or production) costs, there always exists a price equilibrium. In this case, each firm is a local monopolist in the sense that, at the equilibrium prices, *firms do not interact* (cross elasticities are zero). In contrast, when the reservation price is "high enough," encroachments on the market areas of neighboring competitors become profitable to the firm, thus making the competitive process more stringent. Not surprisingly, the existence problem now becomes very tricky. An extensive analysis of this important issue is provided in 4.1.

4. SPATIAL OLIGOPOLISTIC COMPETITION

In this section, we apply the concepts introduced above to spatial competition with a *given* number of firms. We have subdivided the section according to the strategic variables controlled by the firms, i.e., price only (4.1), location only (4.2) and, price and location (4.3).

4.1. Variable prices and parametric locations

Let us go back for a moment to the examples treated in 3.3 where we assume that π is high enough for the two firms to interact. The first example shows that, because of the mutually exclusive choices made by the consumers, discontinuities in demand functions and nonexistence of a price equilibrium should be the rule in the case of an atomic distribution of consumers. Accordingly, we shall not consider this representation anymore, and shall focus in the sequel on models embodying a nonatomic distribution of consumers, as in our second example. Given that the existence of a price equilibrium is restored in this example, the following question suggests itself: Is the assumption of a nonatomic distribution of consumers sufficient to guarantee the existence of a price equilibrium? Clearly priority must be given to this problem since, obviously, no property of the price equilibrium can be fruitfully investigated before elucidating the conditions under which equilibrium exists.

In solving existence problems, it is usual to rely on fixed-point arguments. These arguments can be applied when payoff functions are quasi-concave.[24] In oligopoly theory, a sufficient condition to ascertain the quasi-concavity of profits is to establish the concavity of demand functions. For a given density Δ of consumers and a given transportation cost function t, we know that the demand function of firm j is given by

$$D_j(p_1, \ldots, p_j, \ldots, p_n) = \int_{A_j(p_1,\ldots,p_n)} \Delta(s)\, ds,$$

where A_j is defined as in 3.2. Thus, properties of the demand functions most crucially hinge on both the functions Δ and t. Accordingly the existence problem amounts to determining the classes of consumer density functions and transportation cost functions giving rise to concave demand functions. Actually, this turns out to be a very difficult problem: even with a continuum of consumers, it is easy to find density functions Δ and transportation cost functions t for which the resulting demand functions are *not* concave.

To illustrate this, let us go back to our second example of 3.3, but let us assume only that the transportation cost function t is increasing and twice continuously differentiable. Denote by $\bar{s}(p_1, p_2)$ the location of the consumer who is indifferent between purchasing from firm 1 at price p_1 and purchasing from firm 2 at price p_2, i.e., $\bar{s}$ is the solution to

$$p_1 + t(\bar{s} - s_1) = p_2 + t(s_2 - \bar{s}).$$

[24] For a discussion of the existence of Nash equilibrium in noncooperative games, see, e.g., Friedman [34].

A standard calculation shows that

$$\frac{d^2\bar{s}}{dp_1^2} = -\frac{t''|_{(\bar{s}-s_1)} - t''|_{(s_2-\bar{s})}}{[t'|_{(\bar{s}-s_1)} + t'|_{(s_2-\bar{s})}]^3}.$$

When $\bar{s}$ is between s_1 and s_2, we obtain $D_1(p_1, p_2) = \bar{s}(p_1, p_2) - s_1$ and $D_2(p_1, p_2) = s_2 - \bar{s}(p_1, p_2)$. The concavity of the function D_1 is equivalent to the concavity of $\bar{s}(p_1, p_2)$ in p_1, so that $d^2\bar{s}/dp_1^2$ should be nonpositive. Let $\bar{p}_2$ be fixed and such that $\bar{p}_2 + t(s_1, s_2) > 2t(s_1, s_2)$. Then, for

$$p_1' \underset{\text{def}}{=} \bar{p}_2 + t(s_1, s_2),$$

we obtain

$$\bar{s}(p_1', \bar{p}_2) = s_1,$$

while for

$$p_1'' \underset{\text{def}}{=} \bar{p}_2 - t(s_1, s_2)$$

we have

$$\bar{s}(p_1'', \bar{p}_2) = s_2.$$

Consequently,

$$\left.\frac{d^2\bar{s}}{dp_1^2}\right|_{(p_1', \bar{p}_2)} = -\left.\frac{d^2\bar{s}}{dp_1^2}\right|_{(p_1'', \bar{p}_2)},$$

so that, unless $d^2\bar{s}/dp_1^2$ equals zero on the interval $[p_1', p_1'']$, $d^2\bar{s}/dp_1^2$ must change its sign on this interval, and $\bar{s}$ cannot be concave on the whole domain $[0, \pi]$. Hence the foregoing shows that, in all location problems on the line for which there exist pairs of prices yielding a market boundary between any two firms, *it makes no sense to assume demand concavity from the outset.* Since, in particular, all models of spatial competition "à la Hotelling" fall in this category, it is hopeless to tackle the corresponding existence problem by the traditional approach relying on concavity.

Of course, concavity of demand is only a sufficient condition for existence. Nevertheless, as illustrated below, a significant departure from concavity may yield nonexistence, even if the profit functions of the firms are continuous.

Assume that the unit of length is chosen so that the total length of the market is equal to 1, and that the transportation cost function is of the linear-quadratic type:[25]

$$t(s', s'') = c\,|s' - s''| + d(s' - s'')^2, \qquad c > 0, \qquad d > 0,$$

for any two locations s' and s'' in $[0, 1]$. Furthermore suppose that both firms are symmetrically located as $s_1 = \frac{1}{2} - a$ and $s_2 = \frac{1}{2} + a$, respectively ($\frac{1}{2} > a > 0$). This situation is illustrated in Figure 8.

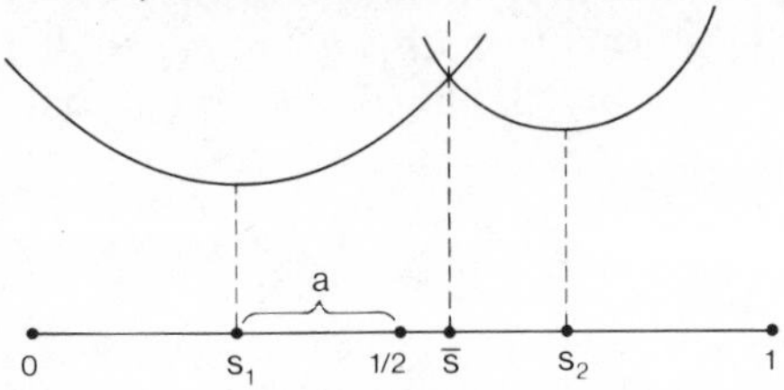

FIGURE 8.

For a fixed value of p_2, $\bar{p}_2$ say, the demand function to firm 1 is easily derived as

$$D_1(p_1, \bar{p}_2)$$

$$= 0, \qquad \text{if } p_1 \geq p_1' \underset{\text{def}}{=} \bar{p}_2 + 2ac + 2ad: \qquad \text{region 1}$$

$$= \frac{\bar{p}_2 - p_1 + 2ad + 2ac}{4ad}, \quad \text{if } p_1' > p_1 \geq p_1'' \underset{\text{def}}{=} \bar{p}_2 + 2ac + 4a^2 d: \quad \text{region 2}$$

$$= \frac{\bar{p}_2 - p_1 + 2ad + c}{4ad + 2c}, \quad \text{if } p_1'' > p_1 \geq p_1''' \underset{\text{def}}{=} \bar{p}_2 - 2ac - 4a^2 d: \quad \text{region 3}$$

$$= \frac{\bar{p}_2 - p_1 + 2ad - 2ac}{4ad}, \quad \text{if } p_1''' > p_1 \geq p_1'''' \underset{\text{def}}{=} \bar{p}_2 - 2ac - 2ad: \quad \text{region 4}$$

$$= 1, \qquad \text{if } p_1'''' > p_1 \geq 0: \qquad \text{region 5.}$$

Thus, the demand function is piecewise linear; it is depicted in Figure 9.

The corresponding price regions are represented in Figure 10.

Notice that, at the price $p_1 = p_1'''$, where the market boundary appears exactly at $s_1 = \frac{1}{2} + a$, the demand function exhibits a kink

[25] Such a transportation cost function can be derived from a quadratic utility function which includes distance in its arguments. It might be argued against the linear-quadratic specification that it does not fit observed average transportation costs which are, first, decreasing and, beyond some threshold, increasing in distance. To meet this objection, it is sufficient to add a fixed component to the linear-quadratic function on account of indivisibilities in departure-and-arrival time. This modification does not affect the structure of demand.

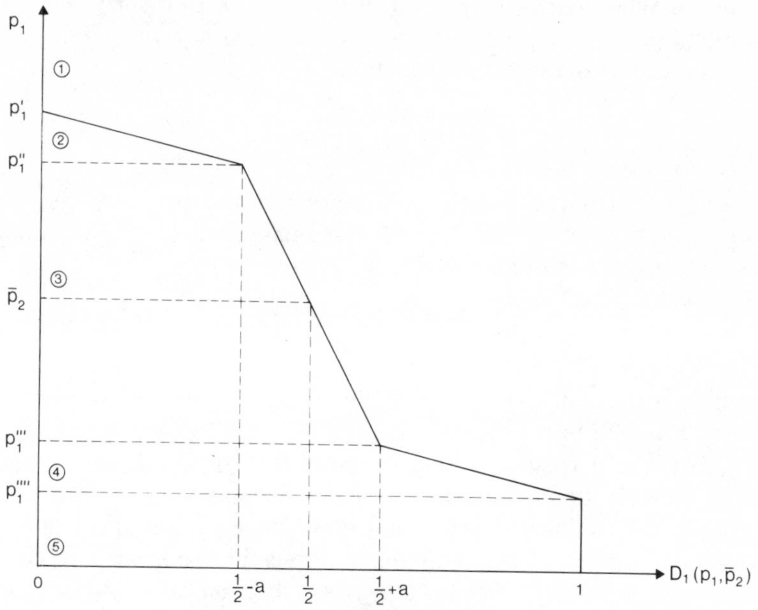

FIGURE 9.

destroying its concavity. Corresponding to the five domains of the demand function, profits exhibit five different configurations. For high prices, the whole market is served by firm 2. Then from $p_1 = p_1'$, profits are given by the quadratic function

$$p_1 \cdot \frac{\bar{p}_2 - p_1 + 2ad + 2ac}{4ad}$$

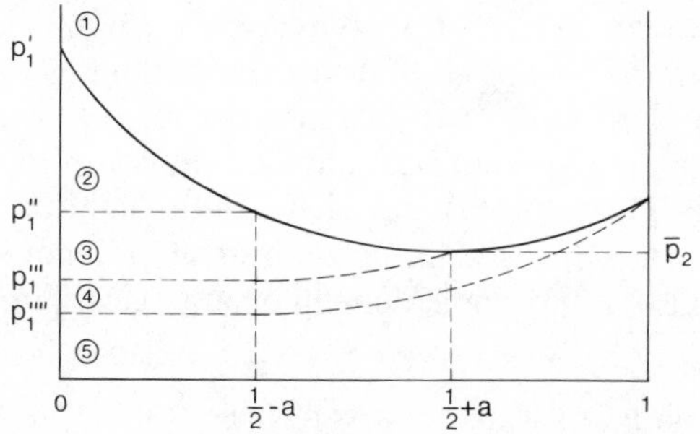

FIGURE 10.

until the value $p_1 = p_1''$ is reached; then the price p_1 is such that the market boundary is precisely at $s_2 = \frac{1}{2} - a$. From now on, profits are on a branch of the quadratic function

$$p_1 \cdot \frac{\bar{p}_2 - p_1 + 2ad + c}{4ad + 2c},$$

as long as the market boundary is in between the two firms, i.e., as long as p_1 remains larger than p_1'''. For smaller values of p_1, profits are given by the quadratic function

$$p_1 \cdot \frac{\bar{p}_2 - p_1 + 2ad - 2ac}{4ad}$$

until p_1 is so small ($<p_1''''$) that total demand is addressed to firm 1; profits are then linear in p_1.

Clearly the particular pattern of the profit function depends both on $\bar{p}_2$, and on the parametric values a (which measures the distance between the two firms) and c and d (which measures the relative weight of the quadratic and linear terms in the transport cost function respectively). Should a price equilibrium exist, the natural candidate for fulfilling the role is the pair (p_1^*, p_2^*), where p_1^* is the best reply against p_2^* in the domain where the market boundary is in-between the two firms, and vice versa. (Indeed, we should not expect equilibrium strategies elsewhere since that would imply that at least one of the two firms does not completely serve its own hinterland. In this case, that firm would have an incentive to lower its price so as to recover it, contradicting the fact that such a pair is a price equilibrium.[26]) To ascertain that the pair (p_1^*, p_2^*) is a price equilibrium, one must further exclude the possibility that a firm can increase its profits by charging a price which would capture customers located *beyond* his competitor's shop. Consider this possibility from the viewpoint of firm 1. Depending on the values of the parameters a, c and d, two cases may arise when firm 2 chooses p_2^*; the corresponding configurations of firm 1's profits $P_1(p_1, p_2^*)$ are depicted in Figures 11(a) and 11(b). In the first case, illustrated by Figure 11(a), $P_1(p_1, p_2^*)$ is maximized at p_1^*, not only in the region 3, but also in the whole domain of strategies. Firm 1 does not

[26] This is formally proved in the first stage of Proposition 1 in the Appendix.

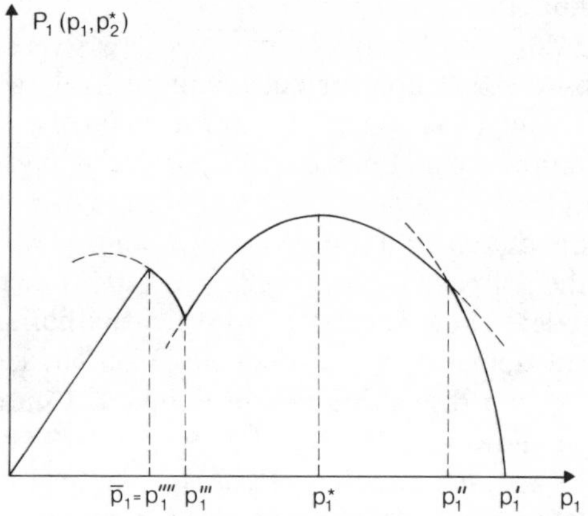

FIGURE 11a.

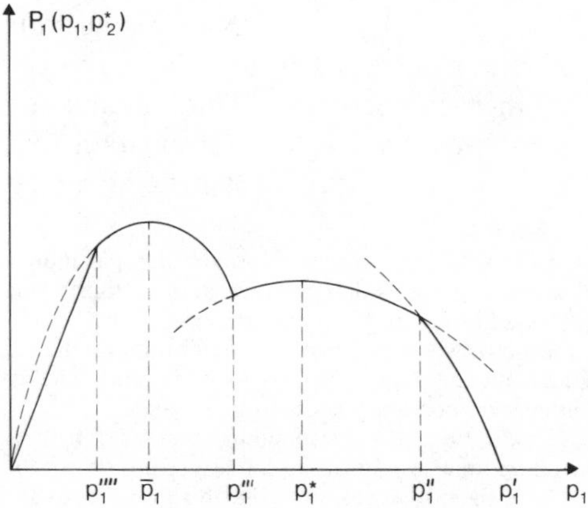

FIGURE 11b.

gain by charging price $\bar{p}_1$ (which yields the highest profit for price strategies leading to a market boundary inside firm 2's hinterland). However, in the second case, as illustrated by Figure 11(b), profits of firm 1 are higher when it uses $\bar{p}_1$ instead of p_1^*: then (p_1^*, p_2^*) cannot be a price equilibrium! As shown in Proposition 1, in the Appendix, this case must necessarily arise when $a < \frac{1}{4}$ and $c/d > 2a/$

$1-4a$, entailing the nonexistence of a price equilibrium in the corresponding domain of parameters.[27]

We thus observe that nonexistence is more likely when firms are close to each other (a is "small"), and/or when the linear term of the transportation cost function is relatively higher than the quadratic term (c/d is "large"). When $c=0$, the demand function is concave (since $d\bar{s}^2/dp_1^2=0$ on $[p_1', p_1'''']$), and existence is guaranteed. In the opposite case, when $d=0$, we fall back on Hotelling's model. Then, for $a<\frac{1}{4}$, no price equilibrium exists.[28] In this case, some domains of the demand function degenerate into points, yielding two discontinuities in demand. Contrary to widespread opinion, however, we see that *the nonexistence of a price equilibrium is not necessarily related to the existence of these discontinuities;* rather it is the non-quasi-concavity of the profit functions which may pose problems.[29]

Various solutions have been proposed to fill the gap created by the absence of an equilibrium in spatial price competition. First, it was suggested by Eaton [23] that a way of coping with this difficulty consists in preventing the firms from using strategies which completely eliminate their competitors.[30] This amounts to restricting the set of strategies to prices yielding nonzero market shares. When the definition of a price equilibrium is amended on account of this

[27] In fact, Proposition 1 does more: it provides a partition of the space of parameters (c/d, a) into those which guarantee, and those which exclude, the existence of a price equilibrium in this particular example.

[28] Very often, the nonexistence of a price equilibrium in the standard Hotelling model is even not mentioned. The reason is to be found in the argument which is followed. First, the market boundary is determined, assuming that it lies in between the two sellers. Second, for the corresponding demand functions, the first-order conditions are solved and the second-order conditions are shown to hold. What is so obtained is therefore a pair of prices such that no firm can increase its profits by changing its price unilateraly, *conditional* upon the assumption that both sellers are in business. However, on account of the equilibrium condition, the possibility must also be recognized to each firm to price its competitor out of business. Actually, when the two sellers are close enough, the undercutting strategy turns out to be more profitable for at least one firm, thus generating instability. See d'Aspremont, Jaskold Gabszewicz and Thisse [13] for a complete treatment of the linear cost case ($d=0$).

[29] Note that using nonuniform distribution of consumers does not prevent the nonexistence of a price equilibrium for some values of the parameters, as shown by Shilony [89].

[30] The intuitive reason for this assumption is as follows: if one firm's change in price results in the exit of its competitor, then the competitor will react immediately by dropping its price. See also Eaton and Lipsey [26] and Novshek [71].

restriction, we obtain the so-called "modified zero conjectural variation" (ZCV) price equilibrium.[31] It is indeed true that, in the case of linear transportation costs, a modified ZCV price equilibrium does exist, whatever the locations of the firms are (see footnote 28). Nevertheless, as shown in Proposition 2 of the Appendix, a sufficient departure from the linear case may also invalidate this statement for exactly the same reason as in the case of a price equilibrium, i.e., the absence of quasi-concavity of the profit functions over the restricted domain of prices.[32]

Another attempt for solving the existence problem amounts to permitting firms to use mixed strategies in pricing. This line of research has been followed by Dasgupta and Maskin [12] who show the existence of an equilibrium in mixed strategies, even with the kind of discontinuities encountered above.[33] Nevertheless, even if mixed strategies have been successfully used in some price competition models (see, e.g., Allen and Hellwig [1] and Varian [96]), we are not convinced by what often looks like a "deus ex-machina" for determining an equilibrium solution. In particular, we are not aware of a justification, in terms of firms' behavior, for the use of mixed strategies in spatial price competition.[34]

---

[31] Formally a modified ZCV price equilibrium is a pair of prices $(\tilde{p}_1, \tilde{p}_2)$ such that $\tilde{p}_1 D_1(\tilde{p}_1, \tilde{p}_2) \geq p_1 D_1(p_1, \tilde{p}_2)$, and $\tilde{p}_2 D_2(\tilde{p}_1, \tilde{p}_2) \geq p_2 D_2(\tilde{p}_1, p_2)$ for all $(p_1, p_2) \in \{(p_1, p_2); D_1(p_1, p_2) > 0,\ D_2(p_1, p_2) > 0\}$.

[32] There is also a conceptual difficulty with the modified ZCV price equilibrium. The restriction imposed to the sets of strategies is especially relevant when firms are close to each other. However, it is precisely in that case that the incentive for undercutting the competitor's price is strong! In some sense, the modified ZCV assumption eliminates one of the basic ingredients of price competition since it artificially excludes by essence any form of price war.

[33] An illustration is provided by Shilony [88] who was able to compute the equilibrium mixed strategies in a specific model of spatial price competition with an atomic distribution of consumers.

[34] For example, a possible justification for the use of mixed strategies is related to the so-called "fictitious play" procedure, defined for infinitely repeated games: at stage t, $t = 1, \ldots, \infty$, player i chooses a pure strategy maximizing his expected payoff using as probability density, on player i's pure strategy choice, the frequency with which player j has used each pure strategy in the past stages of the game. If the game is zero-sum, and if both players act as described by this procedure, then the sequence of frequency distributions converges to the pair of mixed strategies corresponding to an equilibrium point of the game. However, in the present context, this justification suffers from two drawbacks. Firstly, it is not clear why the firms should behave according to the assumption of the fictitious play, even if the game is infinitely repeated. Secondly, and more importantly, the above-mentioned result ceases to be true for the non-zero-sum games (for a counter-example, see Shapley [87]).

Is there some hope to restore existence of an equilibrium when the "geography" of the model is no longer represented by a linear segment? An alternate representation, suggested by Samuelson [81], takes the circle as the geographical basis along which consumers and firms are distributed. Unfortunately, as shown in Proposition 3 of the Appendix, the difficulty encountered above comes up again when firms are sufficiently close to each other, i.e., as soon as the two firms are established within 90° of each other.[35]

Common to both the linear and circular representations is that the market boundary may lie in-between the two firms.

By contrast, this is not possible when firms are located *outside* the segment where consumers are themselves located.[36] To capture this idea we consider the following modification of the previous example. Assume again that the consumers are evenly distributed

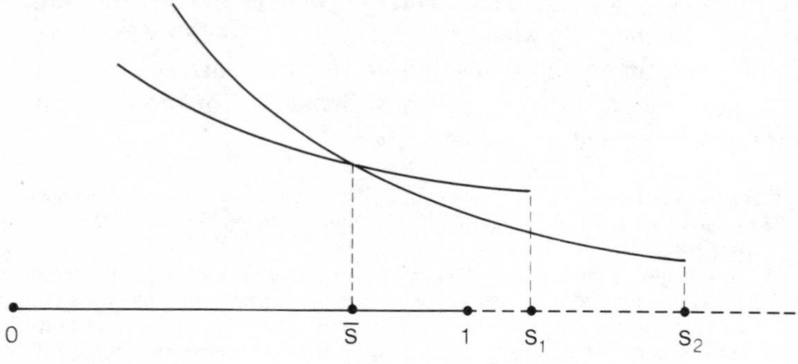

FIGURE 12.

[35] When both firms are diametrically opposed on the circle, there exists a price equilibrium. This seems to corroborate the idea that there would be an equilibrium when transportation costs are linear and firms are equally spaced along the circle, a well-established model in location theory. (See, for instance, Lancaster [62], Novshek [71], Salop [80]). Indeed, it is easy to show that, in this case, a price equilibrium does exist, *whatever the number of firms*. Nevertheless, no general argument is available for a broad class of transportation cost functions. Worse than that, it can also be shown that demand functions are not concave in prices in most other cases. Thus, as above, the extension of the existence property here also poses problems.

[36] For an intuitive interpretation of this model, think of supermarkets which must be located outside a residential area spread along a road. See also de Meza and von Ungern-Sternberg [20].

along the interval [0, 1], but that both firms are now established on the right side of this interval, with firm 1 located at $s_1 \geqslant 1$ and firm 2 at $s_2 \geqslant s_1$, as in Figure 12 (contrast with Figure 8). As above, assume that $t(s', s'') = c\,|s' - s''| + d(s' - s'')^2$ for any pair of locations (s', s'').

For a fixed value of p_2, $\bar{p}_2$ say, the demand function addressed to firm 1 is easily derived as

$$
\begin{aligned}
D_1(p_1, \bar{p}_2) &= 0, && \text{if } p_1 \geqslant p_1' \underset{\text{def}}{=} \bar{p}_2 + c(s_2 - s_1) + d(s_2^2 - s_1^2);\\
&= \frac{\bar{p}_2 - p_1}{2d(s_2 - s_1)} + \frac{c + d(s_2 + s_1)}{2d}, && \text{if } p_1' > p_1 \geqslant p_1'' \underset{\text{def}}{=} \bar{p}_2 + (c - 2d)(s_2 - s_1) + d(s^2 - s^2);\\
&= 1, && \text{if } p_1'' > p_1 \geqslant 0.
\end{aligned}
$$

Proposition 4 in the Appendix shows that, in this reformulation of our duopoly example where the firms are located *outside* the segment in which the consumers are located, a price equilibrium always exists; nevertheless, if the firms are far apart from each other, the firm located farthest from the consumers has a zero market share at the corresponding equilibrium. These results are to be contrasted with those obtained when firms are established *within* the consumers' area. There we have seen that for a wide domain of parameters, no price equilibrium exists while when it does exist, both firms must serve a positive fraction of total demand. Surprisingly, it seems that more stability in noncooperative price behavior is to be expected when one of the two firms is endowed with a strict exogenous advantage over the other one, as in the case where both firms are located outside the consumers' area. The fact that, in this case, seller 1's location is viewed as strictly better by all consumers than seller 2's location, prevents the latter from using price strategies which would attract to him the whole market. This privilege is reserved to firm 1: in fact, if the two firms are located sufficiently far apart, firm 1 uses this advantage to quote an equilibrium price which leaves no market share to firm 2, even with the latter quoting a zero price. This asymmetry between sellers no longer exists when the market boundary may lie between them, i.e., when shops are located inside the consumers' area. In that case, both firms may exert the privilege of undercutting their competitor, leading possibly to price instability.

Interestingly, the product differentiation counterpart of the "geographical" advantage of firm 1 over firm 2 discussed above is the phenomenon where one firm sells a product unanimously ranked higher by all consumers than some other substitute sold by another firm. In the other case, when no firm has a geographical advantage over its competitor, this unanimity is lost in the ranking by consumers of the differentiated substitutes. This is reminiscent of the distinction introduced by Lancaster [62] between *vertical* and *horizontal* product differentiation, respectively. Reinterpretation of the above examples within the more general framework of product differentiation suggests that more stability in noncooperative price behavior must be expected under vertical, than under horizontal differentiation.[37]

Let us now briefly summarize the above discussion, and draw some general conclusions about the existence of a price equilibrium. It is standard procedure to obtain existence results by relying on fixed-points arguments. Usually, these arguments are restricted to quasi-concave profit functions. Concavity of demand functions which, in turn, implies the quasi-concavity of profits, hinges on the particular properties of the consumer location density function and the transportation cost function. It was thus natural to start our inquiry about existence by studying the class of spatial competition problems giving rise to the required properties. This inquiry was rather disappointing! We have shown that, even in the simplest case of two firms and a uniform density of consumers located along the line, not only may a price equilibrium fail to exist for some reasonable transportation cost functions, but demand concavity can almost never hold if there are pairs of prices yielding a market

[37] This is confirmed in an alternative model of product differentiation when consumers are differentiated by their income levels, instead of locations, and have to choose between two products, one of which is unanimously preferred by all consumers to the other; see Jaskold Gabszewicz and Thisse [54].

boundary between the firms.[38,39] Still more disappointing, the remedies proposed for escaping from the nonexistence problem are not very satisfactory. The modified ZCV price equilibrium need not exist either; and using the circle as a geographical support for consumers and firms does not imply "good behavior" of demand functions. Finally the only positive result is obtained when firms are located outside the consumers' area—thus leading to a unanimous ranking of them by the consumers: in that case, a price equilibrium exists. This is interesting to the extent that the model finds a natural interpretation in product differentiation theory.

This summary shows that *general existence results can hardly be expected in spatial price competition.* Different attitudes are then conceivable. Firstly, one could list the cases for which a price equilibrium exists or does not exist. Such a list would allow us to identify the situations where a noncooperative price equilibrium can be viewed as a reasonable outcome of the market process. It would also identify the cases where alternative types of behavior must be expected from the sellers. In particular, it seems that some form of coordination among firms is called for to stabilize competition when no price equilibrium exists. Secondly, one might think of enriching the model with new ingredients which appear as especially relevant in spatial competition. This is discussed in Section 6.

4.2. Variable locations and parametric prices

In some industries firms do not exert any control over their price level because of either cartel agreements or price administration by public authorities. Such constraints on prices drive competition among firms to alternative paths. In location theory, it is usual to consider that, instead of reducing price in order to attract customers (as under price competition), firms compete in locating their sales outlets so as to guarantee for themselves the largest possible sales.

The notions of industry and competition introduced in 3.1 and 3.2

[38] Remember, however, that a price equilibrium always exists when the reservation price is low enough compared to the transportation costs: it then corresponds to the monopoly price; see 3.3.

[39] Note also that MacLeod [66] has obtained a nonexistence theorem in the n-dimensional case when (i) marginal production costs fall with output; or (ii) firms choose simultaneously prices and capacity levels.

to deal with price choice, can be easily reformulated to deal with location choice. However, a major difference with price competition is observed: if firms are free to choose their location in space S, any firm $j \in N$ can become the direct competitor of any other firm $k \in N$. Indeed, for this to happen, it suffices that the former chooses a location s_j close enough to the location s_k of the latter, to guarantee that their demands become dependent on both s_j and s_k. As a result, we may expect the process of competition in locations to be very harsh.

Another basic ingredient of this type of competition is the existence of discontinuities in the demands addressed to the firms. Indeed, in general, the leapfrogging of one firm by another gives rise to discontinuities in their demands, whatever the transportation cost functions and the consumer density function. To see this, let us assume that two firms with a single outlet each compete along a linear market of unit length. Given the location of firm 1 at $s_1 < \frac{1}{2}$, the size of the market area of firm 2 established at $s_2 = s_1 - \varepsilon$, with $\varepsilon > 0$ arbitrary small, is equal to $s_1 - \varepsilon/2$. Let now firm 2 be located at $s_2' = s_1 + \varepsilon$. The size of its market area is then given by $1 - s_1 - \varepsilon/2$, which is always larger than $s_1 - \varepsilon/2$ since $s_1 < \frac{1}{2}$. In other words, any "crossing" of firms outside the center $\frac{1}{2}$ generates a discontinuity in the sizes of market areas and, therefore, in the demands.

The foregoing suggests that no general result can be reasonably expected in locational competition. However, in some particular, but meaningful, cases it is possible to obtain interesting properties. Let us start again with the simplest case where each firm is allowed to operate only one outlet. Then it is readily apparent that the market share of a firm is the measure of the set of consumers who are located closer to that firm than to any other one. In this model, for a given set of firms, a strategy for a firm is defined as the choice of a location, the payoff to a firm is its market share, and equilibrium—call it a *location equilibrium*—as a configuration of locations such that no single firm prefers an alternative location, given the location of its competitors. Assume first that there are two firms competing along a line of length one with consumers evenly distributed along the line. Let firm 1, say, be located outside the center. In this case, firm 2 can maximize its market share by establishing itself near to firm 1 on the larger side of the market.

But then firm 1 has an incentive to leapfrog its competitor since this allows it to increase its market share. Such behavior prevents any pair of noncentral locations from being an equilibrium. On the other hand, when both firms are placed at the center, each obtains half of the market and any unilateral move by a firm away from the center is accompanied by a decrease in its market share. In other words, *the only location equilibrium consists of both firms clustered at the center of the market.*[40] Note that an equilibrium does exist with two firms, despite discontinuities in the payoff functions. However the existence property does not carry over when the number of firms increases to 3. There, it is clear that both peripheral firms tend to "sandwich" the central firm which finds itself with a vanishing market. As a result, this firm will leapfrog one of its competitors, thus generating instability.[41] For more than three firms, the existence of a location equilibrium is restored. On the other hand, the bunching of firms at a single location does not occur. Instead, we obtain equilibrium configurations with some firms being pairwise located and others being single. Clearly, such equilibria entail some degree of inefficiency since the efficient pattern of firms is here given by the uniformly dispersed locations. For a more detailed analysis of the n-firm case, the reader is referred to Eaton and Lipsey.[42]

We now analyze the more realistic case where, again with given prices, each firm is allowed to multiply its sales outlets. Nonprice competition takes the alternative form of spreading out these sales outlets over the market. However this "proliferation" is limited by the existence of fixed costs associated with the establishment of a new outlet. The resulting trade-off yields a market situation where, given the strategies of its competitors, no firm wants either to

---

[40] This result has come to be known as the "Principle of Minimum Differentiation." For a critical appraisal, see Eaton and Lipsey [25] and Graitson [37].

[41] See Lerner and Singer [63] for more details. Shaked [85] has computed an equilibrium in mixed strategies for the 3-firm case.

[42] Hotelling [50] has suggested reinterpreting the model of spatial competition with given prices for explaining the choice of political platforms in party competition. This idea has been elaborated by Downs [21] and developed further by many others, including Davis, Hinich and Ordeshook [15], Denzau, Kats and Slutsky [18] and Kramer [59].

relocate its outlets, or to modify their number. This situation is labeled below as an "outlet selection equilibrium" at given price $\bar{p}$.

Consider as above that two firms compete to attract a population of identical consumers evenly distributed along a line of unit length. But now each firm is allowed to open an arbitrary number of sales outlets. Without loss of generality, we take $\bar{p}=1$ so that the fixed cost, denoted f, must satisfy $0<f\leqslant\frac{1}{2}$ for the duopoly to be sustainable. Formally, we define an *outlet selection equilibrium* as a pair of integers (m^*, n^*) and a pair of location vectors $(\mathbf{s}_1^*, \mathbf{s}_2^*)$, with $\mathbf{s}_1^*=(s_{11}^*, \ldots, s_{1m^*}^*)$ and $\mathbf{s}_2^*=(s_{21}^*, \ldots, s_{2n^*}^*)$ such that, for all $m\in N_+$ and all location vectors $\mathbf{s}_1=(s_{11}, \ldots, s_{1m})$,

$$P_1(\mathbf{s}_1^*, \mathbf{s}_2^*)=D_1(\mathbf{s}_1^*, \mathbf{s}_2^*)-m^*f\geqslant P_1(\mathbf{s}_1, \mathbf{s}_2^*)=D_1(\mathbf{s}_1, \mathbf{s}_2^*)-mf$$

and, for all $n\in N_+$ and all location vectors $\mathbf{s}_2=(s_{21}, \ldots, s_{2n})$,

$$P_2(\mathbf{s}_1^*, \mathbf{s}_2^*)=D_2(\mathbf{s}_1^*, \mathbf{s}_2^*)-n^*f\geqslant P_2(\mathbf{s}_1^*, \mathbf{s}_2)=D_2(\mathbf{s}_1^*, \mathbf{s}_2)-nf.$$ [43]

Let us develop an intuitive argument showing that, when $1/2f$ is an integer,[44] a unique outlet selection equilibrium exists such that the outlets of the two firms are pairwise located and equally spaced. First, due to the symmetry of the problem, we normally expect both firms to have the same number n of outlets at equilibrium. Furthermore, for any equal number of outlets, the process of competition between the two firms must lead to a configuration where outlets are paired at points which guarantee a market share equal to $1/2n$ to each outlet. Clearly only points $(2i-1)/2n$, $i=1, \ldots, n$, satisfy this property. Now, if $n>1/2f$, each firm has an incentive to increase its number of outlets. Indeed, by "sandwiching" any existing outlet of its competitor, it can attract an additional share of the market equal to $1/2n$ which exceeds f. On the other hand, if $n>1/2f$, then both firms make negative profits, a contradiction to the conditions of equilibrium. Finally, for $n=1/2f$,

[43] As above, $D_i(\mathbf{s}_1, \mathbf{s}_2)$ denotes the demand to firm i at (given) price $\bar{p}=1$, when firm 1 owns m outlets located at $s_{11}, \ldots, s_{1m}$, and firm 2 owns n outlets located at $s_{21}, \ldots, s_{2n}$, respectively. Formally, the demand D_1 is given by the measure of the set of consumers who are closer to the nearest outlet of firm 1 than to the nearest outlet of firm 2; and $D_2=1-D_1$. By convention, when several outlets are located at the same point, the local market is equally shared.

[44] If $1/2f$ is not an integer, then there exists no outlet selection equilibrium. However, provided that f is not too large relative to the market size, $1/2f$ can be well approximated by an integer in which case some approximate equilibrium exists.

any addition or deletion of outlets leaves the profit of each firm unchanged. Therefore, $n^* = 1/2f$ together with the locations $s_{1i}^* = s_{2i}^* = (2i - 1)/2n^*$ is the only outlet selection equilibrium. (A formal proof is given in Proposition 5.)[45]

Some remarks are now in order. First, we observe that, at the outlet selection equilibrium, profits of both firms are driven down to zero. This is the analog, in nonprice competition, of the standard Bertrand outcome under price competition. However, contrary to the Bertrand case where the market solution proves to be socially optimal, the outlet selection equilibrium leads to a locational pattern which does not minimize total (production plus transportation) costs in the economy. To see this, we first notice that the equilibrium number of outlets is not related to the transportation cost function t. It depends only upon the sizes of the market and fixed costs, whereas the socially optimal configuration also depends on the transportation cost function. In addition, every outlet is "duplicated" at the outlet selection equilibrium, while they are all separated at the socially optimal configuration. Note that this could suggest that the equilibrium number—$2n^*$—of outlets is too large. Actually, it can only be said that the equilibrium locations are nonoptimal. Whether or not $2n^*$ is larger or smaller than the socially optimal number of outlets depends on the specific transportation costs.

Second, the outlet selection equilibrium can be compared to the locational configuration resulting from competition among $2n^*$ independent firms running a single outlet each. Using the results of Eaton and Lipsey [25], we see that there exists a unique location equilibrium with each of $2n^*$ firms covering its fixed costs. At this equilibrium firms are paired at points $(2i - 1)/2n^*$, for $i = 1, \ldots, n^*$. Consequently, we see that the two configurations are identical, although $2n^*$ firms operate in the latter case and only 2 in the former.

4.3. Variable prices and locations

In the above discussion, firms were assumed to control a single strategy, i.e., price or location, respectively. In the first case, the

---

[45] It is worth noting that the above argument only requires the assumption that the transportation cost t is an increasing function of distance.

purpose was to deal with spatial price competition among firms already established in the market. The second case was intended to cover spatial competition among firms bound by price agreements or regulation. Now we consider the more general case when the above restrictions are removed. Firms are then allowed to choose *both* price and location. Here firms control their potential market, and, through both their price and location choices, build their network of potential direct and indirect competitors. This choice can be modelled in two different ways. In the first one, firms *simultaneously* choose price and location (simultaneous game) while, in the second case, prices and locations are chosen according to a *two stage process* (sequential game). Again we specialize the analysis to the 2-firm case.

In the simultaneous game, a strategy for firm i, $i=1, 2$, is defined as a pair (p_i, s_i), and a *price-location equilibrium* as a pair of strategies (p_i^*, s_i^*) such that, $\forall p_i \geq 0$, $\forall s_i \in S$, $i=1, 2$, $i \neq j$:

$$P_i((p_i^*, s_i^*), (p_j^*, s_j^*)) \underset{\text{def}}{=} p_i^* D_i((p_i^*, s_i^*), (p_j^*, s_j^*))$$

$$\geq p_i D_i((p_i, s_i), (p_j^*, s_j^*)) \underset{\text{def}}{=} P_i((p_i, s_i), (p_j^*, s_j^*)),$$

where D_i, the demand to firm i, is given by the measure of the set of consumers who patronize firm i. It is folklore that, regardless of the transportation cost function, *no price-location equilibrium can exist in the simultaneous game model.* To see this, assume that a price-location equilibrium exists. Clearly, at this equilibrium, both firms must earn positive profits which implies that $p_1^* \neq 0$ and $p_2^* \neq 0$. Two cases may then arise. In the first one, we have $s_1^* \neq s_2^*$. Without loss of generality, we may assume that firm 2's profits exceed or equal firm 1's profits. Then, firm 1 can increase its profits by locating at $\bar{s}_1 = s_2^*$ and by charging a price $\bar{p}_1 = p_2^* - \varepsilon$, with $\varepsilon > 0$ arbitrarily small. Indeed,

$$P_1((\bar{p}_1, \bar{s}_1), (p_2^*, s_2^*)) > P_2((p_1^*, s_1^*), (p_2^*, s_2^*))$$

since firm 1 now captures the whole market at price $p_2^* - \varepsilon$. As

$$P_2((p_1^*, s_1^*), (p_2^*, s_2^*)) \geq P_1((p_1^*, s_1^*), (p_2^*, s_2^*)),$$

we thus obtain

$$P_1((\bar{p}_1, \bar{s}_1), (p_2^*, s_2^*)) > P_1((p_1^*, s_1^*), (p_2^*, s_2^*)),$$

a contradiction. In the second case, $s_1^* = s_2^*$. Hence we must have $p_1^* = p_2^* \neq 0$, for both firms to have positive profits. But then, by an argument à la Bertrand, it is clear that each firm has an incentive to undercut its competitor, a contradiction.[46]

Since the simultaneous game approach looks like a blind alley, let us turn to the second formulation in terms of a sequential game. There price and location strategies are assumed to be played one at a time in a two-stage process. The division into stages is motivated by the fact that the choice about location is prior to the decision on price. Thus locations are chosen in the first stage and prices in the second stage. Assuming that prices are chosen as in 4.1, at the price equilibrium of the corresponding subgame (if it exists!), the resulting second-stage profits depend only upon the location choice made in the first stage. Accordingly, these profits can be used as payoff functions in the first-stage subgame in which strategies are locations. Now we proceed to a formal definition of an equilibrium concept for this sequential game set-up. Define $(p_1^*(s_1, s_2), p_2^*(s_1, s_2))$ as the price equilibrium corresponding to a location pair (s_1, s_2). Then a *perfect price-location equilibrium* is a pair $((p_1^*, s_1^*), (p_2^*, s_2^*))$ such that:[47]

i) $p_1^* = p_1^*(s_1^*, s_2^*)$;

ii) $p_2^* = p_2^*(s_1^*, s_2^*)$;

iii) $$p_i^*(s_i^*, s_j^*)D_i(s_i^*, s_j^*; p_i^*(s_i^*, s_j^*), p_j^*(s_i^*, s_j^*)) \geq p_i^*(s_i, s_j^*)D_i(s_i, s_j^*; p_i^*(s_i, s_j^*), p_j^*(s_i, s_j^*)),$$

where $D_i(s_i, s_j; p_i^*(s_i, s_j), p_j^*(s_i, s_j))$ denotes the demand to firm i at the price equilibrium corresponding to locations (s_i, s_j). For all we know, Hotelling [50] was the first to use (implicitly) the concept of perfect price-location equilibrium. Indeed, in his duopoly locational analysis, he first solves for prices, given locations, and then, introducing the corresponding equilibrium prices into the profit

[46] Notice that the notion of "modified ZCV price-location equilibrium" can also be introduced in the above simultaneous game, by forbidding firms from undercutting their competitors. This permits one to prove the existence of a modified ZCV price-location equilibrium when transportation costs are linear in distance and when the reservation price is not "too" high; see Kohlberg and Novshek [58].

[47] The rigorous treatment of the subgame perfect (Nash) equilibrium concept is due to Selten [84].

functions, he solves for locations. The concept of perfect equilibrium captures the idea that, when firms choose their locations, they both anticipate the consequences of their choice on price competition. In particular, they are aware that this competition will be more severe if they locate close to each other, rather than far apart. Unfortunately, this concept is meaningful only if, for any location choices by firms, there exists one, and only one, corresponding price equilibrium—otherwise, either payoffs would be undefined or multivalued. From 4.1 we know how demanding these existence, and a fortiori uniqueness, conditions are. Nonetheless, in contrast to the simultaneous game where equilibrium *never* exists, there are various contexts of spatial competition in which the above conditions are met and for which a perfect price-location equilibrium therefore exists.

To illustrate the above concept, let us consider the reformulation of Hotelling's location model with quadratic transportation costs, i.e., $t(s', s'') = d(s' - s'')^2$, $s, s' \in S = [0, 1]$, $d > 0$. It is straightforward to show that firms' demands are given, respectively, by

$$
\begin{aligned}
D_1(p_1, \bar{p}_2) &= 0, && \text{if} \quad p_1 \geqslant p_1' \underset{\text{def}}{=} \bar{p}_2 + d(s_2^2 - s_1^2);\\
&= \frac{\bar{p}_2 - p_1 + d(s_2^2 - s_1^2)}{2d(s_2 - s_1)}, && \text{if} \quad p_1' > p_1 \geqslant p_1'' \underset{\text{def}}{=} \bar{p}_2 - d(s_2 - s_1)\\
& && \qquad (2 - s_1 - s_2);\\
&= 1, && \text{if} \quad p_1 < p_1'';
\end{aligned}
$$

and

$$
\begin{aligned}
D_2(\bar{p}_1, p_2) &= 0, && \text{if} \quad p_2 \geqslant p_2' = \bar{p}_1\\
& && \qquad + d(s_2 - s_1)\\
& && \qquad (2 - s_1 - s_2);\\
&= \frac{\bar{p}_1 - p_2 + d(s_2 - s_1)(2 - s_1 - s_2)}{2d(s_2 - s_1)}, && \text{if} \quad p_2' > p_2 \geqslant p_2''\\
& && \qquad \underset{\text{def}}{=} \bar{p}_1 - d(s_2^2 - s_1^2);\\
&= 1, && \text{if} \quad p_2 < p_2''.
\end{aligned}
$$

It is easily seen that, for *any* location pair (s_1, s_2), the profit functions are quasi-concave in prices, which ensures the existence of a corresponding price equilibrium $(p_1^*(s_1, s_2), p_2^*(s_1, s_2))$. Furthermore, first-order necessary conditions immediately yield

$$p_1^*(s_1, s_2) = d(s_2 - s_1)\left(\frac{2 + s_1 + s_2}{3}\right),$$

$$p_2^*(s_1, s_2) = d(s_2 - s_1)\left(\frac{4 - s_1 - s_2}{3}\right),$$

which is the unique price equilibrium for (s_1, s_2). We then substitute those prices into the demand functions of both firms, and multiplying them by $p_1^*(s_1, s_2)$ and $p_2^*(s_1, s_2)$, respectively, profits are seen to be functions of s_1 and s_2 only. Some routine calculations show that regardless of the location of the other firm the profits of firm 1 decrease when s_1 increases whereas profits of firm 2 increase with s_2. Consequently, each firm gains from moving away as far as possible from its competitor so that the perfect price-location equilibrium is given by

$$((p_1^*, s_1^*), (p_2^*, s_2^*)) = ((d, 0), (d, 1)).$$

The notion of perfect price-location equilibrium is convenient for discussing the behavior of firms in terms of location choice when the impact of their location decisions on prices is explicitly taken into account. In particular it may be used to evaluate the tendency of the firms to cluster or, on the contrary, to separate.[48] Here we notice that, in the modified version of Hotelling's model with quadratic costs, firms definitely strive to move away as far as possible from each other, so that, at the perfect equilibrium, firms locate at the endpoints of the linear market. In the language of product differentiation, this conclusion reveals a tendency for the firms to increase the differences between their products, rather than to minimize them, so as to create submarkets in which each enjoys some degree of monopoly power. Thus, in the case of quadratic costs, the perfect price-location equilibrium, far from confirming Hotelling's statement that clustering should be the rule, implies maximal differentiation.

Would it be possible, however, to observe the clustering of both

---

[48] If prices are invariant, then we know from 4.2. that the clustering of the two firms is the rule.

firms at the perfect price-location equilibrium for *some* particular transportation cost function? The answer is *no*. Indeed when firms settle at the same place s, with $s \in S$, the unique price equilibrium is given by the Bertrand solution, i.e., $p_1^*(s, s) = p_2^*(s, s) = 0$. Consequently the corresponding profits are necessarily zero. On the other hand, for any pair of *distinct* locations, the equilibrium prices are strictly positive (if they exist!), and the corresponding market shares are nonempty. Thus, relocating away from s must increase the profits of each firm, a contradiction.[49] In other words, in a perfect price-location equilibrium, *firms relax price competition by locating apart from each other.*[50,51]

5. SPATIAL COMPETITION WITH FREE ENTRY

In the definition of an industry introduced in 3.3, the number of firms is exogenously given. However a comprehensive definition of an industry should include an *endogenous* determination of that number. In this section we review the most popular approaches to this problem, namely: spatial monopolistic competition (5.1) and sequential entry (5.2).

In spatial monopolistic competition, the long-run equilibrium is viewed as the outcome of a process of entry with portable firms. Firms locate one after each other according to some given order of entry.[52] Furthermore, each time a new firm enters, the incumbents can costlessly re-locate to a new pattern. In other words, history does not matter.

By contrast, in models of sequential entry, firms are supposed to be immobile. When a new firm enters, the incumbents are now committed to their locations; however, they are still free to change

---

[49] A more formal argument can be found in d'Aspremont, Jaskold Gabszewicz and Thisse [14].

[50] This property seems to contradict the "Principle of Minimum Differentiation" suggested by Hotelling [50]. Actually, in the original model (linear transportation costs), nothing can be said about the tendency of both sellers to agglomerate since we know from 4.1 that no price equilibrium exists when they are close to each other. See also Economides [29].

[51] Note that a result by Gal-Or [35] for the case of linear transportation costs suggests a similar tendency for the firms to separate when the price game is solved in mixed strategies.

[52] For a multiple firm entry dynamics, see Eaton [24] and Grace [36].

their prices. Consequently, the long-run equilibrium does depend on the sequence of decisions taken by the firms.[53]

In both types of model, it is assumed that entry stops when the profits of a potential entrant evaluated at the post-entry market solution are no longer positive.[54]

In this section, the simplifying assumptions made in Section 4 are maintained throughout.

5.1. Spatial monopolistic competition

The "Monopolistic Competition Revolution" (this term was coined by Samuelson [81]) is closely associated with the work of Chamberlin [10]. Basically this "revolution" centers about the following market conditions: (i) the industry is made up of a large number of firms, each selling one product only; (ii) products are differentiated in some manner, so that each firm faces its own demand function; (iii) the entry of a new firm (or product) has only a negligible effect on the individual demand functions of the incumbents: this amounts to assuming that the new firm captures a few customers only from each one of them; (iv) entry proceeds until profits in the industry are driven down to zero. Conditions (i) and (ii) are perfectly typified by the process of spatial competition for an homogeneous product. Indeed if firms are free to enter and to choose their location in space, we should expect many firms, each of them with its own circle of customers. Now what about conditions (iii) and (iv)? Actually condition (iv) is less of a problem,[55] but the realism of condition (iii) turns out to be questionable. This has not escaped the scrutiny of Kaldor [56] in his remarkable appraisal of Chamberlin's work. Indeed he was the first to point out that a new

[53] An alternative interpretation is that the price-location equilibrium of the spatial monopolistic competition model is the result of a "tâtonnement" process of entry, in which firms do not settle down before the equilibrium number is reached. On the other hand, the configuration produced by a sequential entry can be viewed as the outcome of a "non-tâtonnement" process of entry.

[54] This assumption of "sophisticated behavior" has been made explicit in the theory of entry-deterrence by excess capacity in Dixit [19] and Spence [90]. Actually, it was already implicit in the modern approaches to spatial competition with free entry; see, e.g., Capozza and Van Order [8].

[55] The zero-profit condition must be replaced by a nonnegative profit condition on account of the integrability of the number of firms.

firm must necessarily be placed in between two existing firms so that its entry will only affect the demand addressed to these neighbors, and not necessarily the demand to firms located farther away from it. Conversely, the demand for the entrant's product will be most sensitive to the prices of these neighbors, but hardly sensitive to the prices of more distant firms.[56] This casts some doubt on the direct applicability of Chamberlin's theory to the realm of spatial competition. Nonetheless, an amended version of it—which has come to be known under the name of spatial monopolistic competition—can be used to build an endogenous determination of the industry size. This amended version combines the idea of using a system of "chained" demands (as suggested by Kaldor) in the determination of noncooperative equilibrium prices (given the number of firms), together with the existence of set-up costs for determining the equilibrium number of firms.

To illustrate this, we consider a slightly modified version of a model due to Salop [80], but inspired from the earlier work of Kaldor [56] and Lösch [64].[57] Assume a circular market of unit length with a uniform density Δ of consumers, and with n firms established at equidistant locations. The establishment of a firm gives rise to a fixed cost denoted by f.[58] Then the profits of a firm can be written as

$$P_i(n) = p_i D_i(p_{i-1}, p_i, p_{i+1}) - f,$$

where D_i, the demand to firm i, is defined as

$$D_i(p_{i-1}, p_i, p_{i+1}) = \left(\frac{p_{i-1} - 2p_i + p_{i+1}}{2d/n} + \frac{1}{n}\right)\Delta$$

in the domain of prices where all firms are in the market (we

---

[56] In the terminology introduced in Section 3, this means that, in some restricted range of prices, neighbors are direct competitors while more distant firms are indirect competitors only.

[57] The model of spatial monopolistic competition has attracted the attention of several authors. Kaldor's and Lösch's ideas have been elaborated by Mills and Lav [69] and developed, among others, by Beckmann [4], Capozza and Van Order [8], Eaton [24], Lancaster [62] and Novshek [71].

[58] In the present context, some degree of increasing returns are necessary for the existence of economic units called firms. Otherwise, production would take place (at an infinitesimal level) in each point of space.

assume again quadratic transportation costs, i.e., $t(s', s'') = d(s' - s'')^2$).

Observe that the demand to firm i decomposes into two terms. The first one, i.e., $((p_{i-1} - 2p_i + p_{i+1})/2d/n)\Delta$, embodies Kaldor's remark that the demand of a firm only depends upon the prices set by its neighbors while the second term, i.e., Δ/n, reflects Chamberlin's idea that increasing the number of firms in the industry lowers demand addressed to that firm.

First consider the problem of finding a price equilibrium when n is parametric ($n > 1$). Applying first-order conditions, we obtain

$$p_{i-1} - 4p_i + p_{i+1} + \frac{2d}{n^2} = 0.$$

The unique solution to this system of difference equations is given by $p_i = d/n^2$, for $i = 1, \ldots, n$. The second-order conditions are trivially satisfied so that this vector of prices is a natural candidate for a price equilibrium. More precisely, it remains to show that no firm can increase its profits by pricing its neighbors out of business. Clearly such is the case since any positive price cannot attract to a firm consumers lying beyond its neighbor's location. Consequently, as long as the corresponding profits $\Delta(d/n^3) - f$ exceed zero, the vector of prices

$$(p_1^*, \ldots, p_i^*, \ldots, p_n^*) = \left(\frac{d}{n^2}, \ldots, \frac{d}{n^2}, \ldots, \frac{d}{n^2}\right)$$

is a price equilibrium.

It is now easy to determine the number of firms resulting from free entry. As long as the individual profit with $n + 1$ firms is nonnegative, i.e., $\Delta(d/(n+1)^3) \geqslant f$, space is left for the entry of an $(n+1)$th firm. Accordingly, spatial monopolistic competition is achieved for the largest integer, say n^*, which satisfies $n \leqslant \sqrt[3]{(d/f)\Delta}$ (the entry-condition) while the corresponding price equilibrium is

$$\left(\frac{d}{n^{*2}}, \ldots, \frac{d}{n^{*2}}, \ldots, \frac{d}{n^{*2}}\right).$$

Substituting $p_{i-1}^* = p_{i+1}^* = d/n^{*2}$ into D_i, yields

$$D_i(p_{i-1}^*, p_i, p_{i+1}^*) = \frac{2\Delta}{n^*} - \frac{n^*\Delta}{d} p_i,$$

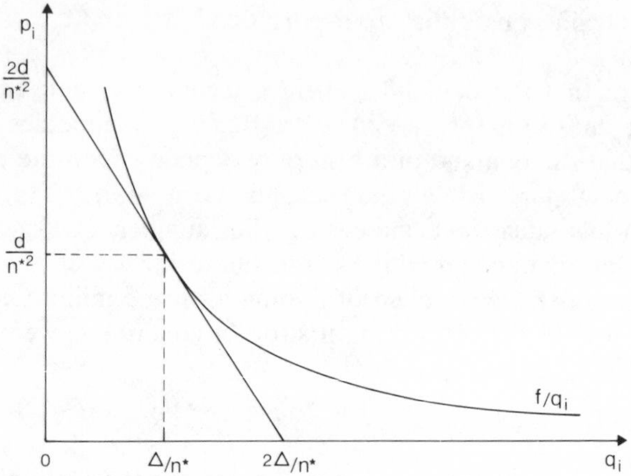

FIGURE 13.

which looks like a Chamberlinian individual demand function, conditional on the fact that all firms, but firm *i*, charge equilibrium prices. Figure 13 illustrates the spatial monopolistic competition solution when the quantity $q_i = \Delta/n^*$ leads exactly to zero profits, i.e., the Chamberlinian tangency condition.

Until now, no explanation has been afforded as to why firms are equally spaced.[59] Clearly, for a given value of *n*, the symmetric pattern is not a simultaneous Nash equilibrium in price and location (see 4.3). Nevertheless, it can be shown to be the outcome of the first stage of a perfect price-location equilibrium (see Eaton and Wooders [28] and Economides [30]). This means that the long-run equilibrium in the industry is viewed as the subgame perfect equilibrium of a three-stage game in which the first stage pertains to the entry decision, the second to the location choice and the third to the price choice.

[59] Most of the contributions to spatial monopolistic competition fail to provide a rigorous treatment of the regular spacing property. The standard argument used in these contributions is as follows. Firstly, the derivatives of the profit function with respect to price and with respect to location are computed. Secondly, it is shown that a regular spacing and a uniform price solve simultaneously the system of first-order conditions and satisfy the second-order conditions. So doing, an equilibrium is obtained in which no firm can increase its profits by a *local* change in its price and/or location. This does not mean, however, that these strategies maximize a firm's profits over the whole domain of price and location. For more details, see Novshek [71].

Inspection of the entry-condition above reveals that the equilibrium number of firms results from a trade-off between transportation costs, which are borne by consumers, and set-up costs, due to the production activity of firms. In the case of low fixed costs, we see that the number of firms in the industry becomes very large. As a result, the equilibrium price approaches zero, i.e., the competitive level.[60] The reason is that several firms are now established in the vicinity of each consumer's location, thus multiplying the number of substitutes and strengthening price competition. On the other hand, if transportation costs are low, only a small number of firms can survive. Yet equilibrium price is still close to the competitive level. In spite of the small number of firms (products), products become almost perfect substitutes since the only reason for differentiation among products lies in the existence of transportation costs. Surprisingly enough, the competitive solution can therefore result from sharply contrasted economic contexts: *prices are close to marginal costs when transportation costs are low and firms are few, as well as when set-up costs are low and firms are many*. In addition, the entry-condition indicates that, ceteris paribus, a higher density Δ of consumers leads to an industry structure with a larger number of firms. Thus, in a "large" economy, we similarly expect the price equilibrium to be close to the competitive level.[61]

Also associated with the theory of monopolistic competition is the so called "excess capacity" theorem: at the market solution, each firm produces at less than minimum average production cost. It is worth noting that the same property holds when we consider the sum of production *and* transportation costs. To show this, let C be the production cost function and T the minimum transportation cost per firm when each firm serves a fraction $1/n$ of the whole market (recall that the length of the circular market has been normalized to one). Total (production *and* transportation) costs are equal to $nC(1/n) + nT(1/n)$, where

$$T\left(\frac{1}{n}\right) = \int_{-1/2n}^{1/2n} \Delta t(x)\, dx,$$

[60] This is reminiscent of Novshek [72].

[61] A detailed discussion of the asymptotic properties of the long-run equilibrium is given by Eaton and Wooders [28] for L-shaped and U-shaped cost functions.

with $t(x)$ as the unit transportation cost function. Treating n as a continuous variable and differentiating this expression with respect to n yields

$$\frac{C(1/n)}{1/n} - C'(1/n) + \frac{T(1/n)}{1/n} - T'(1/n) = 0.$$

Assuming that $t(x)$ is increasing with distance, we see that $T(1/n)/(1/n)$, i.e., the average transportation cost, is smaller than the transportation cost for serving the marginal consumer, namely $T'(1/n)$. Thus, it follows that $C(1/n)/(1/n) > C'(1/n)$, namely, *the average production cost exceeds the marginal production cost at the number of firms minimizing the sum of production and transportation costs.* The possibility of observing "excess capacity" arising at the market solution should therefore be evaluated by reference to the level of production per firm minimizing production plus transportation costs, and *not* by reference to the level of production per firm minimizing production costs only. Figure 14 illustrates this property, where $\bar{q}$ and $\bar{\bar{q}}$ denotes the quantities minimizing average production plus transportation costs, and average production costs, respectively.

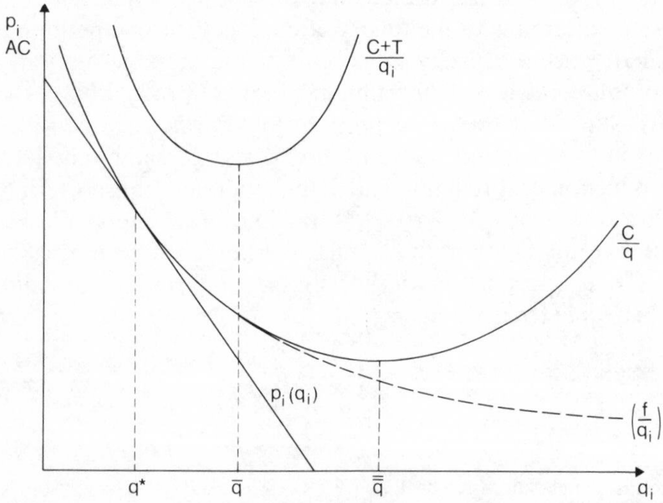

FIGURE 14.

It remains to compare the market solution $q^* = \Delta/n^*$ with $\bar{q}$. In the case of the above example, it is straightforward to show that $\bar{n}$, corresponding to the number of firms minimizing production plus transportation costs, is the largest integer for which the condition $n \leqslant \sqrt[3]{(d/6f)\Delta}$ holds. Accordingly we have $q^* < \bar{q} = \Delta/\bar{n}$, as in Figure 14, and there is still "excess capacity." Nevertheless nothing guarantees that, for alternative transportation cost functions t, spatial monopolistic competition would lead to a number of firms n^* for which Δ/n^* would fall in between $\bar{q}$ and $\bar{\bar{q}}$. In that case, the Chamberlinian excess capacity theorem would be invalidated, and spatial monopolistic competition would involve "too few" firms in the industry.[62]

A current claim in spatial monopolistic competition is that, since firms are mobile, profits should be driven down to zero at the long-run equilibrium. In the following, we show by means of an example that this claim is not true: portable firms may earn substantial pure profits at the long-run equilibrium.

Consider a linear market of unit length with a unit density of consumers; transportation costs are of the linear-quadratic type. At the initial stage, a single firm is located at $s_1 = 1$. Clearly, for a reservation price π large enough, the whole market is served at the monopoly price $\pi - c - d$. Now assume that a firm considers the possibility of entry, but that entry is restricted to occur to the right of s_1 (as in Figure 11), i.e., $s_2 \geqslant s_1 = 1$. A quick look at Proposition 4 in the Appendix reveals that, for $c/d \geqslant 4 - s_1 - s_2$, the price equilibrium corresponding to locations s_1 and $s_2 \geqslant s_1$, is given by $p^*(s_1, s_2) = (s_2 - s_1) \cdot [d(s_1 + s_2 - 2) - c]$ and $p_2^*(s_1, s_2) = 0$. Thus, if $c \geqslant 2d$, the condition $c/d \geqslant 4 - s_1 - s_2$ is always satisfied and the profits of the potential entrant are unavoidably equal to zero. Consequently, entry is deterred even though the existing firm may earn substantial pure profits. Interestingly enough, this property is not related here to the existence of indivisibilities in production.

[62] Note that the excess capacity theorem is known to be invalid in the following case. Let the parameters of the model be such that the configuration maximizing the social surplus is made of a single firm. When fixed costs are large enough, the maximum profit level achievable by a monopolist is always negative. Accordingly, no profit-maximizing firm will serve the market, even though optimality requires a positive production. As pointed out by Heal [47], this "insufficient capacity" result can be explained by the fact that firms charge mill pricing; under delivered pricing, the distorsion vanishes (see 2.1).

Instead, it rests on the nature of preferences: all consumers prefer firm 1 to firm 2 when both firms charge the same price.[63]

In spite of its attractiveness, the model of spatial monopolistic competition suffers from a serious drawback: the existence property does not extend to a broad class of transportation cost functions. For instance, with linear transportation costs, the symmetric circular pattern is no longer a perfect equilibrium. We know indeed that a price equilibrium does not exist for some asymmetric locations (see 4.1) so that the perfect equilibrium concept becomes meaningless. In such cases, the interpretation of the model becomes very problematic.

5.2. Sequential entry

In the real world, one often observes that location decisions can hardly be modified whereas prices are relatively flexible. At the limit, location decisions can be considered as irrevocable while prices can be changed ad libitum and instantaneously. If firms do not enter simultaneously but, instead, sequentially, this entails an asymmetry in the choices open to the incumbents and to the entrant, respectively. Whereas the former have to stick to their existing locations, though they are still free to choose their price, the latter enjoys the freedom of choice in both price and location.

Let us illustrate this asymmetry in the following simple example. Assume that a single firm is located at s_1^* on a circular market with unit length and unit consumer density. Imagine that a second firm considers entry. If it anticipates that post-entry prices will be set at the price equilibrium with two firms, the entrant cannot do better than choosing a location which maximizes its profits evaluated at the corresponding equilibrium prices. We know that in the case of quadratic transportation costs, this best location is diametrically opposed to s_1^*, i.e., at s_2^* as shown in Figure 15. The resulting equilibrium prices are $p_1^* = p_2^* = d/4$, while profits are equal to $d/8 - f$. Suppose that $f < d/8$, so that entry occurs. Assume now that a third firm is contemplating entry. Again, if the potential

[63] Reinterpreted within the framework of vertical product differentiation, this property is equivalent to the "finiteness' property on the number of firms which can cohabit simultaneously at equilibrium, as shown by Jaskold Gabszewicz and Thisse [55] and extended by Shaked and Sutton [86].

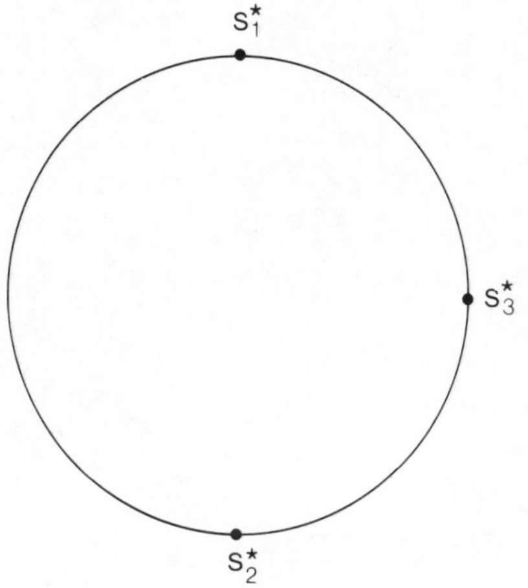

FIGURE 15.

entrant anticipates the new price equilibrium obtained in the case of three firms, it will choose a location maximizing profits evaluated at post-entry equilibrium prices. The profit-maximizing location for this firm can easily be shown to be situated at any middle point s_3^* between the incumbents locations, as in Figure 15. The post-entry equilibrium prices are respectively $p_1^* = p_2^* = \frac{14}{128}d$ while $p_3^* = \frac{11}{128}d$, thus generating profits equal to $\frac{495}{4096}d$ for firms 1 and 2, and $\frac{121}{4096}d$ for the potential entrant. We immediately notice the asymmetry between the incumbent's revenues due to both lower price and smaller market share for the entrant. This asymmetry can be so strong that the entrant's revenues are not sufficient to cover his set-up costs: then entry does not occur. More specifically, when $f > \frac{121}{4096}d$, only the existing two firms can survive.

It is worth noticing that the immobility of the incumbents may be essential for the long-run determination of the number of firms in the industry. Indeed, in the context of the above example, if the existing two firms were able to costlessly re-locate, the entrant will anticipate that prices and locations are adjusted at the perfect price-location equilibrium with three firms, as illustrated in Figure 16. Both price and market share of the entrant are now higher. Accordingly, it may happen that *entry is profitable in the portable*

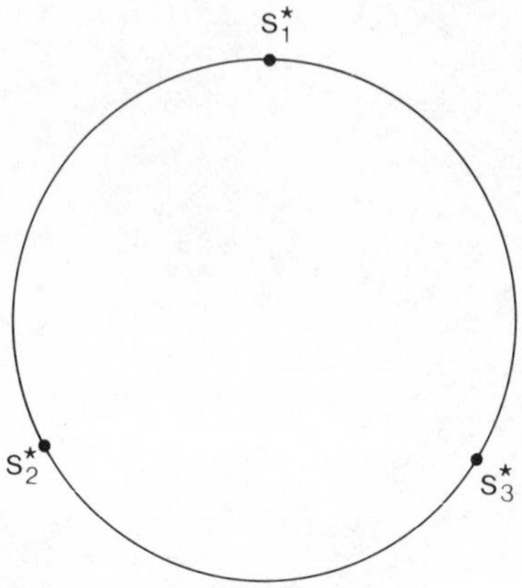

FIGURE 16.

case while *it is not in the immobile case* (this occurs when $\frac{121}{4096}d < f < d/27$). This means that the immobility of firms, paired with sequential entry, is another reason for the persistence of substantial pure profits at the long-run equilibrium.[64,65]

When entry is sequential and when location decisions are made once-for-all, it seems reasonable to expect that an entrant also anticipates subsequent entry by *future* competitors. Accordingly, at each stage of the entry process, the entrant must consider as given the locations of firms entered at earlier stages but can treat the locations of firms entering at later stages as conditional upon his own choice. In other words, the entrant is a follower with respect to the incumbents, and a leader with respect to future competitors. The location chosen by each entrant is then obtained by backward induction, from the optimal solution of the location problem faced by the ultimate entrant, to the entrant himself. This is the essence of the solution concept proposed independently by Hay [46], Prescott and Visscher [77] and Rothschild [78].

To give an intuitive feeling for the concepts involved, let us go back to the example just described above but in which prices are given and normalized to one. We assume that fixed costs permit the

[64] This was first pointed out by Eaton and Lipsey [26].

[65] As observed by Capozza and Van Order [9], entry barriers in the form of immobile capital may improve efficiency when "too many" firms are involved at the spatial monopolistic competition equilibrium.

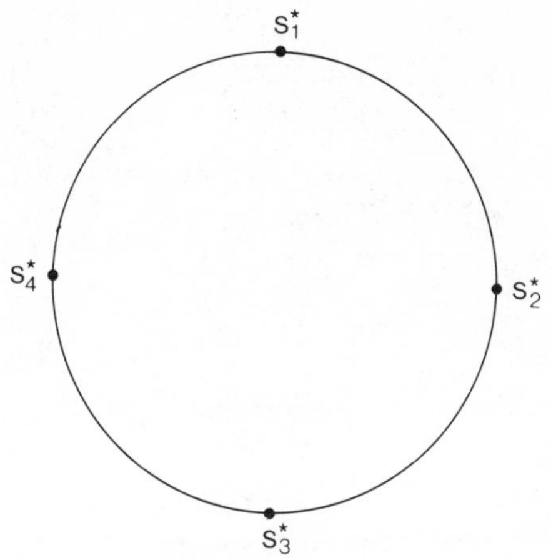

FIGURE 17.

survival of only four firms, if symmetrically located (i.e., $\frac{1}{5} < f \leq \frac{1}{4}$). Given any location $\bar{s}_1$ for firm 1, what could be an "optimal" location for firm 2, knowing that firms 3 and 4 are on the waiting list? If firm 2 establishes itself at a diametrically opposed position—a short-sight strategy—entry of firms 3 and 4 as in Figure 17 are clearly possible.

Making the decisions of firms 3 and 4 conditional on its own, firm 2 can do much better. Indeed if, instead of locating at the antipodal position, firm 2 locates at a distance $2f - \varepsilon$ of $\bar{s}_1$, with $\varepsilon > 0$ arbitrarily small, it succeeds in deterring entry of firm 4. Given the locations $\bar{s}_1$ and $\bar{s}_2$ of firms 1 and 2, firm 3's best choice is to locate at $\frac{1}{2} + f - \varepsilon/2$ (note the asymmetry between firms 1 and 2), so that the largest market share left to firm 4 (namely $f - \varepsilon/2$) is never sufficient to cover its set-up costs. An illustration is given in Figure 18 for $f = \frac{1}{4}$.

Thus, the main contribution of this approach is to reveal that entry can be *deterred* if firms are sophisticated enough to take full advantage of their position in the order of entry.[66] Of course, firms

[66] If firms are allowed to operate several outlets, the first entrant's advantage is dramatically reinforced. Indeed, he always has the possibility of pre-empting the market by establishing several outlets, thus preventing the entry of new comers, as argued by Eaton and Lipsey [27]. See also Schmalensee [82]. The extent to which this policy can indeed be implemented remains an open question.

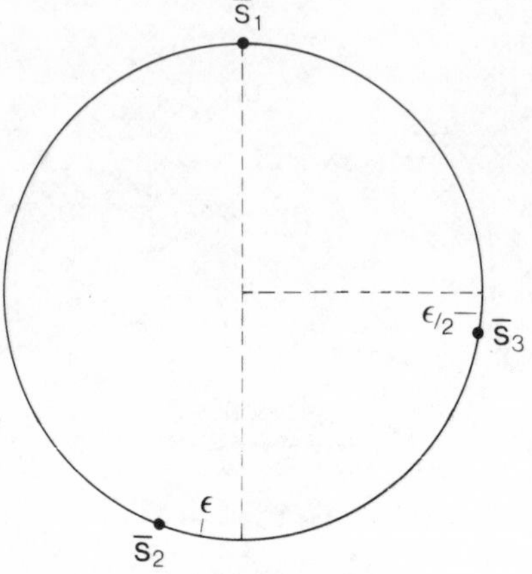

FIGURE 18.

are no longer symmetrically located and pure profits are preserved at the resulting long-run equilibrium.

6. REFORMULATIONS AND CONCLUSIONS

Throughout this article, we have repeatedly encountered the problem of nonexistence of a price and/or location equilibrium. In this last section, we briefly discuss two recent reformulations of the basic model which aim at solving this problem. The first one involves altering the assumption that consumers necessarily buy from the firm with the lowest full price. The second one supposes that firms bear the cost of transporting the product and set discriminatory prices.[67]

(i) At the root of several nonexistence results lies the standard assumption that consumers patronize the cheapest firm. Now,

[67] Other reformulations which might be of interest deal with (i) firms setting quantities instead of prices (Brander [7] and Greenhut and Greenhut [38]), firms selling differentiated products to imperfectly informed consumers (Stahl [92] and Stuart [94]). Nevertheless, it must be said that these contributions did not intend to solve the existence problem in itself. Hence it is not yet clear how much the above-mentioned reformulations can help in restoring the existence of equilibrium in spatial competition.

empirical evidence supports the idea that consumers also take variables other than full price into account. Because of the unobservability of these variables, firms can at best determine the shopping behavior of a particular consumer up to a probability distribution. More precisely, following de Palma *et al.* [17], it is now assumed that firms model the utility of a consumer at s and purchasing from firm i at s_i as a random variable

$$u_{is} = -p_i - t(s, s_i) + \varepsilon_i,$$

where ε_i is a random variable with zero mean. Since consumers maximize utility, firms then evaluate the probability P_{is} that a consumer at s will purchase from firm i as

$$P_{is} = Pr\left[u_{is} = \max_{j=1,\ldots,n} u_{js}\right].$$

If we assume that the variables ε_i are identically, independently Weibull-distributed, then we obtain the logit model (see McFadden [68]), i.e.,

$$P_{is} = \frac{e^{-[p_i+t(s,s_i)]/\sigma}}{\sum_{j=1}^{n} e^{-[p_j+t(s,s_j)]/\sigma}},$$

where σ is the standard-deviation of ε (up to a multiplicative constant).[68] The function P_{is} is depicted in Figure 19 for different values of σ in the case of 2 firms selling at the same given price.

For a bounded market of unit length, a uniform distribution of consumers and a linear transportation cost function, the (expected) demand to firm i is given by

$$D_i = \int_0^1 \frac{e^{-(p_i+c\,|s-s_i|)/\sigma}}{\sum_{j=1}^{n} e^{-(p_j+c\,|s-s_j|)/\sigma}}\, ds.$$

The following remarks are in order. First, for any positive value of the standard deviation, demand D_i is a continuous function of prices $(p_1, \ldots, p_n)$ and of locations $(s_1, \ldots, s_n)$, while D_i is known to be discontinuous when $\sigma = 0$.[69] In particular, there is no discontinuity

[68] It is worth noting that, for $\sigma \to 0$, the limit of P_{is} is equal to 1 when $p_i + t(s, s_i) = \mathrm{Min}_{k=1}^{n} \{p_k + t(s, s_k)\}$, and 0 otherwise. In this case, the above model reduces to the standard one.

[69] Note that the continuity property still holds in the case of an atomic distribution of consumers. Contrast with 3.2.

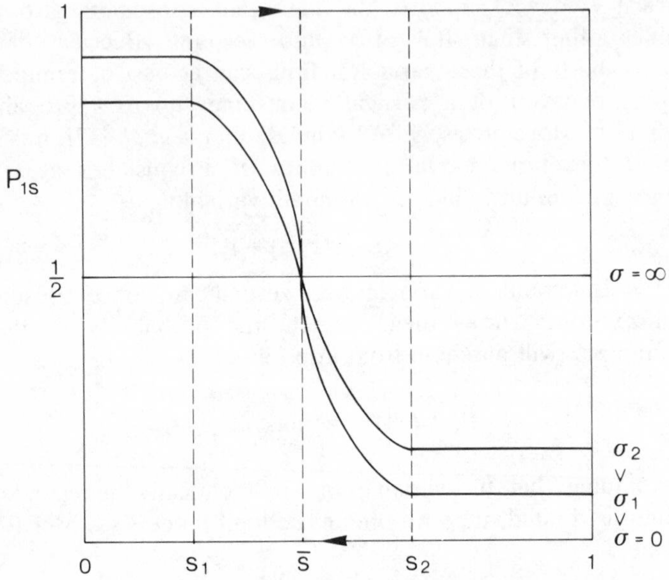

FIGURE 19.

in demand when two firms "cross each other." Second, for any $\sigma > 0$, the set of consumers is no longer segmented into disjoint market areas. Instead, we have a system of completely overlapping market areas. The reason is that the demand addressed to a firm now depends on the prices set by all firms in the industry. Nevertheless, we still observe some "partial" chain effect: mutatis mutandis, the impact of a price change by firm j on firm i's demand decreases as the number of firms located between i and j increases.

The results obtained by de Palma *et al.* [17] suggest that this reformulation of Hotelling's model partially solves the existence problem. First, in the case of parametric prices, they show that for $\sigma/c \geq 1 - 2/n$, there exists a location equilibrium with all firms clustered at the center of the market.[70] Second, in the simultaneous price and location game, the existence of an equilibrium is guaranteed for $\sigma/c \geq 1$; at the price-location equilibrium, firms are located at the market center and charge the same positive price

[70] Note that, for $n = 2$, the clustering occurs for any σ nonnegative; see 4.2.

given by $\sigma n/(n-1)$. In particular, this implies that *the principle of minimum differentiation holds when the standard deviation of the random term is large enough*; and that competition among firms established at the center of the market does not drive prices down to the competitive level (unlike Bertrand), thus yielding positive profits. Quite obviously, these results are in sharp contrast to those obtained in 4.2 and 4.3.

(ii) In the standard model of spatial price competition, consumers travel to the firms and each firm sets a single (mill) price. Instead, when firms deliver their products, each firm can choose a *price function* which specifies the (delivered) price $p(s)$ to be charged to the consumers located at s. The Hotelling problem under delivered pricing is therefore much less constrained than the initial version. As a result, the existence of an equilibrium can be expected to be a less problematic issue.[71]

To see this, assume that two firms, producing at a constant and uniform marginal cost c, are located at s_1 and s_2 in $[0, 1]$ respectively. These two firms compete in (delivered) price on each local market $s \in [0, 1]$ separately. In spite of their identical production costs, they act under asymmetric cost conditions. Indeed, at each $s \in [0, 1]$, $s \neq (s_1 + s_2)/2$, the firms bear different transportation costs. Consequently, over

$$S_i = \{s \in [0, 1];\ t(s_i, s) < t(s_j, s)\},$$

firm i benefits from a cost advantage relative to firm j which allows it to undercut any price set by its rival at $s \in S_i$. This implies that the undercutting process will stop when firm j can no longer reduce its price, i.e., when it is as low as its marginal production and transportation cost $c + t(s_j, s)$. The equilibrium price of firm i at $s \in S_i$ is therefore given by the marginal cost of firm j at s;[72] on the other hand, the equilibrium price of firm i at any point of $[0, 1] - S_i$ is equal to its own marginal cost at this point. A similar argument applies to firm j.

[71] No price equilibrium exists under uniform delivered pricing; see Schuler and Hobbs [83]. A solution in mixed strategies is given in Beckmann [5].

[72] By convention, consumers $s \in S_i$ are supposed to buy at price $c + t(s_j, s)$ from firm i and not from firm j (see 3.2). An alternative view gives the equilibrium price of firm i as $c + t(s_j, s) - \varepsilon$, with $\varepsilon > 0$ arbitrarily small.

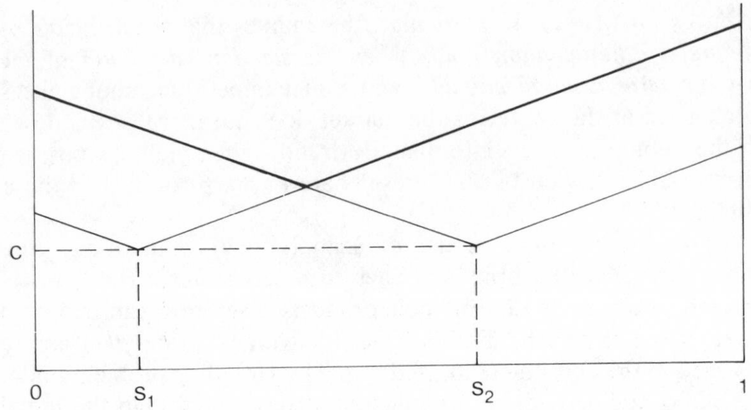

FIGURE 20.

Furthermore, when $s_1 = s_2$, the two firms compete under symmetric cost conditions so that the equilibrium price at s corresponds to the marginal cost at s, i.e., the Bertrand solution.

In conclusion, for any pair of locations s_1 and s_2, *there exists a unique "price schedule" equilibrium* given by

$$\max\{c + t(s_1, s), c + t(s_2, s)\}$$

(see Hoover [48] and Hurter and Lederer [52]).[73] An example is depicted in Figure 20 for the case of linear transportation costs (the equilibrium price schedule is in heavy lines).

It is worth noting that the existence property holds for any continuous transportation cost function. Moreover, the price schedule of firm i decreases with distance as shown in Figure 20, an example of what has come to be known under the name of freight absorption (see, e.g. Phlips [76]).

Let us now come to location choice. Recently, Hurter and Lederer [52] have proved that a perfect price (schedule)-location equilibrium exists in the above model. At the equilibrium, each firm is located at a point which minimizes total transportation costs over its market area, S_i. (Compare this with the similar result obtained in

[73] Note that the existence of a price schedule equilibrium also holds with an atomic distribution of customers. Contrast with 3.3.

2.1.) This implies, first, that firms do not agglomerate (nor locate at the endpoints of the market) and, second, that the two locations minimizing total transportation costs—corresponding to the locations chosen by a two-outlet monopolist—are equilibrium locations. Thus, here also, the results obtained differ significantly from those discussed in 4.1 and 4.3. This clearly indicates that different spatial price policies lead to different locational choices.[74]

Our approach in this survey, treading in Hotelling's footsteps, has been to use location theory to formulate precisely some issues that economic theory can no longer neglect. Among them are the following: What is an industry? What are the price and quantity outcomes in industries where conscious interaction among firms is explicitly recognized? How does nonprice competition operate among firms acting under price agreements? How does entry lead to an endogenous determination of industry size and firms' profits?

We have written this article because we are convinced that the field of spatial competition is more central to economic theory than its present peripheral status would suggest. Hopefully, at the end of this survey, the reader will share our belief that location theory can be viewed as the prototype model of imperfect competition, and that most of the main questions related to this field can be meaningfully and precisely addressed within its framework.

We have selected a limited number of topics and, for each of them, we have confined ourselves to the fundamentals. No doubt this selection process has caused us to neglect some important aspects of the theory. As a result, numerous location theorists who have contributed to the field might feel frustrated by our choice of menu: we owe our apologies to them for not having given full account of their work.

References

[1] Allen, B. and M. Hellwig, "Bertrand-Edgeworth Oligopoly in Large Markets," *Review of Economic Studies,* forthcoming.

[2] Archibald, G. C., B. C. Eaton and R. G. Lipsey, "Address Models of Value Theory," in *New Developments in the Analysis of Market Structure,* ed. by J. Stiglitz and F. Matthewson. Cambridge (Mass.): The MIT Press, 1985.

[74] A similar observation was already made about the spatial monopoly; see Section 2.

[3] Aumann, R. J., "Markets with a Continuum of Traders," *Econometrica,* **32** (1964), 39–50.
[4] Beckmann, M. J., "Spatial Cournot Oligopoly," *Papers of the Regional Science Association,* **28** (1972), 37–47.
[5] Beckmann, M. J., "Spatial Oligopoly as a Noncooperative Game," *International Journal of Game Theory,* **2** (1973), 263–268.
[6] Bertrand, J., "Théorie mathématique de la richesse sociale," *Journal des Savants,* **48** (1883), 499–508.
[7] Brander, J. A., "Intra-Industry Trade in Identical Commodities," *Journal of International Economics,* **11** (1981), 1–14.
[8] Capozza, D. R. and R. Van Order, "A Generalized Model of Spatial Competition," *American Economic Review,* **68** (1978), 896–908.
[9] Capozza, D. R. and R. Van Order, "Unique Equilibria, Pure Profits, and Efficiency in Location Models," *American Economic Review,* **70** (1980), 1046–1053.
[10] Chamberlin, E. H., *The Theory of Monopolistic Competition.* Cambridge (Mass.): Harvard University Press, 1933.
[11] Cornuejols, G., G. L. Nemhauser and L. A. Wolsey, "The Uncapacitated Facility Location Problem," in *Discrete Location Theory,* ed. by R. L. Francis and P. Mirchandani. New York: Wiley-Interscience, 1986.
[12] Dasgupta, P. and E. Maskin, "The Existence of Equilibrium in Discontinuous Economic Games, 1: Theory," *Review of Economic Studies,* forthcoming.
[13] d'Aspremont, C., J. Jaskold Gabszewicz and J.-F. Thisse, "On Hotelling's Stability in Competition," *Econometrica,* **47** (1979), 1045–1050.
[14] d'Aspremont, C., J. Jaskold Gabszewicz and J.-F. Thisse, "Product Differences and Prices," *Economics Letters,* **11** (1983), 19–23.
[15] Davis, O. A., M. J. Hinich and P. C. Ordeshook, "An Expository Development of a Mathematical Model of the Electoral Process," *American Political Science Review,* **64** (1970), 426–448.
[16] Dearing, P. M. and R. L. Francis, "A Minimax Location Problem on a Network," *Transportation Science,* **8** (1974), 333–343.
[17] de Palma, A., V. Ginsburgh, Y. Y. Papageorgiou and J.-F. Thisse, "The Principle of Minimum Differentiation Holds Under Sufficient Heterogeneity," *Econometrica,* **53** (1985), 767–781.
[18] Denzau, A., A. Kats and S. Slutsky, "Multi-Agent Equilibria With Market Shares and Ranking Objectives," *Social Choice and Welfare,* **2** (1985), 95–117.
[19] Dixit, A. K., "The Role of Investment in Entry Deterrence," *Economic Journal,* **90** (1980), 95–106.
[20] de Meza, D. and T. von Ungern-Sternberg, "Monopoly, Product Diversity and Welfare," *Regional Science and Urban Economics,* **12** (1982), 313–324.
[21] Downs, A., *An Economic Theory of Democracy.* New-York: Harper & Row, 1957.
[22] Durier, R. and C. Michelot, "Geometrical Properties of the Fermat–Weber Problem," *European Journal of Operational Research,* **20** (1985), 332–343.
[23] Eaton, B. C., "Spatial Competition Revisited," *Canadian Journal of Economics,* **5** (1972), 268–278.
[24] Eaton, B. C., "Free Entry in One-Dimensional Models: Pure Profits and Multiple Equilibria," *Journal of Regional Science,* **16** (1976), 21–33.
[25] Eaton, B. C. and R. G. Lipsey, "The Principle of Minimum Differentiation Reconsidered: Some New Developments in the Theory of Spatial Competition," *Review of Economic Studies,* **42** (1975), 27–49.

[26] Eaton, B. C. and R. G. Lipsey, "Freedom of Entry and the Existence of Pure Profits," *Economic Journal,* **88** (1978), 455–469.
[27] Eaton, B. C. and R. G. Lipsey, "The Theory of Market Pre-emption: The Persistence of Excess Capacity and Monopoly in Growing Spatial Markets," *Economica,* **46** (1979), 149–158.
[28] Eaton, B. C. and M. H. Wooders, "Sophisticated Entry in a Model of Spatial Competition," *The Rand Journal of Economics,* **16** (1985), 282–297.
[29] Economides, N. S., "The Principle of Minimum Differentiation Revisited," *European Economic Review,* **24** (1984), 345–368.
[30] Economides, N. S., "Symmetric Equilibrium Existence and Optimality in Differentiated Product Markets," University of Columbia, Department of Economics, Discussion Paper 197, 1983.
[31] Edgeworth, F., "The Pure Theory of Monopoly," in *Papers Relating to Political Economy.* London: Macmillan, 1925.
[32] Erlenkotter, D., "A Dual-Based Procedure for Uncapacitated Facility Location," *Operations Research,* **26** (1978), 992–1009.
[33] Friedman, J. W., "Reaction Functions and the Theory of Duopoly," *Review of Economic Studies,* **35** (1968), 257–272.
[34] Friedman, J. W., *Oligopoly and the Theory of Games.* Amsterdam: North-Holland, 1977.
[35] Gal-Or, E., "Hotelling's Spatial Competition as a Model of Sales," *Economics Letters,* **9** (1982), 1–6.
[36] Grace, S. H., "Professor Samuelson on Free Enterprise and Economic Efficiency: A Comment," *Quarterly Journal of Economics,* **84** (1970), 337–345.
[37] Graitson, D., "Spatial Competition à la Hotelling: A Selective Survey," *Journal of Industrial Economics,* **31** (1982), 13–25.
[38] Greenhut, J. and M. L. Greenhut, "Spatial Price Discrimination, Competition and Locational Effects," *Economica,* **42** (1975), 401–419.
[39] Greenhut, M. L., *Plant Location in Theory and Practice.* Chapel Hill (N.C.): University of North Carolina Press, 1956.
[40] Greenhut, M. L., *A Theory of the Firm in Economic Space.* Austin: Lone Star Publisher, 1971.
[41] Gronberg, T. and J. Meyer, "Transport Inefficiency and the Choice of Spatial Pricing Mode," *Journal of Regional Science,* **21** (1981), 541–549.
[42] Guelicher, H., "Einige Eigenschaften optimaler Standorte in Verkehrsnetzen," *Schriften des Vereins für Sozialpolitik,* Neue Folge, **42** (1965), 111–137.
[43] Hakimi, S. L., "Optimum Location of Switching Centers and the Absolute Centers and Medians of a Graph," *Operations Research,* **12** (1964), 450–459.
[44] Handler, G. Y. and P. B. Mirchandani, *Location on Networks.* Cambridge (Mass.): The MIT Press, 1979.
[45] Hanjoul, P. and J.-F. Thisse, "The Location of a Firm on a Network," in *Applied Decision Analysis and Economic Behaviour,* ed. by A. J. Hughes Hallet. Den Haag: Martinus Nijhoff, 1984.
[46] Hay, D. A., "Sequential Entry and Entry-Deterring Strategies," *Oxford Economic Papers,* **28** (1976), 240–257.
[47] Heal, G., "Spatial Structure in the Retail Trade: A Study in Product Differentiation With Increasing Returns," *Bell Journal of Economics,* **11** (1980), 565–583.
[48] Hoover, E. M., "Spatial Price Discrimination," *Review of Economic Studies,* **4** (1937), 182–191.

[49] Hoover, E. M., *The Location of Economic Activity*. New York: McGraw-Hill, 1948.

[50] Hotelling, H., "Stability in Competition," *Economic Journal,* **39** (1929), 41–57.

[51] Huriot, J.-M. and J. Perreur, "Modèles de localisation et distance rectilinéaire," *Revue d'Economie Politique,* **83** (1973), 640–662.

[52] Hurter, A. P. and P. J. Lederer, "Spatial Duopoly with Discriminatory Pricing," *Regional Science and Urban Economics,* **15** (1985), 541–553.

[53] Isard, W., *Location and Space-Economy*. New York: John Wiley and Sons, 1956.

[54] Jaskold Gabszewicz, J. and J.-F. Thisse, "Price Competition, Quality and Income Disparities," *Journal of Economic Theory,* **20** (1979), 340–359.

[55] Jaskold Gabszewicz, J. and J.-F. Thisse, "Entry (and Exit) in a Differentiated Industry," *Journal of Economic Theory,* **22** (1980), 327–338.

[56] Kaldor, N., "Market Imperfection and Excess Capacity," *Economica,* **2** (1935), 35–50.

[57] Kats, M. L., "Multiplant Monopoly in a Spatial Market," *Bell Journal of Economics,* **11** (1980), 519–535.

[58] Kohlberg, E. and W. Novshek, "Equilibrium in a Simple Price-Location Model," *Economics Letters,* **9** (1982), 7–15.

[59] Kramer, G. H., "A Dynamical Model of Political Equilibrium," *Journal of Economic Theory,* **16** (1977), 310–334.

[60] Krarup, J. and P. M. Pruzan, "The Simple Plant Location Problem: Survey and Synthesis," *European Journal of Operational Research,* **12** (1981), 36–81.

[61] Kuhn, H. W., "On a Pair of Dual Nonlinear Programs," in *Nonlinear Programming,* ed. by J. Abadie. New York: John Wiley and Sons, 1967.

[62] Lancaster, K., *Variety, Equity and Efficiency*. Oxford: Basil Blackwell, 1979.

[63] Lerner, A. and H. W. Singer, "Some Notes on Duopoly and Spatial Competition," *Journal of Political Economy,* **45** (1937), 145–186.

[64] Lösch, A., *Die räumliche Ordnung der Wirtschaft*. Jena: Gustav Fisher, 1940. English translation: *The Economics of Location*. New Haven (Conn.): Yale University Press, 1954.

[65] Love, R. F. and J. G. Morris, "Mathematical Models of Road Travel Distances," *Management Science,* **25** (1979), 130–139.

[66] MacLeod, W. B., "On the Non-existence of Equilibria in Differentiated Product Models," *Regional Science and Urban Economics,* **15** (1985), 245–262.

[67] Manne, A. S., "Plant Location Under Economies of Scale-Decentralization and Computation," *Management Science,* **11** (1964), 213–235.

[68] McFadden, D., "Conditional Logit Analysis of Qualitative Choice Behavior," in *Frontiers in Econometrics,* ed. by P. Zarembka. New York: Academic Press, 1973.

[69] Mills, E. S. and M. R. Lav, "A Model of Market Areas With Free Entry," *Journal of Political Economy,* **72** (1964), 278–288.

[70] Norman, G., "Pricing System, Distribution of Demand, and Location," *Regional Studies,* **48** (1977), 183–189.

[71] Novshek, W., "Equilibrium in Simple Spatial (or Differentiated Product) Models," *Journal of Economic Theory,* **22** (1980), 313–326.

[72] Novshek, W., "Cournot Equilibrium With Free Entry," *Review of Economic Studies,* **47** (1980), 473–486.

[73] Novshek, W. and H. Sonnenschein, "Marginal Consumers and Neoclassical Demand Theory," *Journal of Political Economy,* **87** (1979), 1368–1376.

[74] Palfrey, T. S., "Spatial Equilibrium With Entry," *Review of Economic Studies,* **51** (1984), 139–156.

[75] Perreur, J. and J.-F. Thisse, "Central Metrics and Optimal Location," *Journal of Regional Science,* **14** (1974), 411–421.

[76] Phlips, L., *The Economics of Price Discrimination.* Cambridge: Cambridge University Press, 1983.

[77] Prescott, E. C. and M. Visscher, "Sequential Location Among Firms With Foresight," *Bell Journal of Economics,* **8** (1977), 378–393.

[78] Rothschild, R., "A Note on the Effect of Sequential Entry on Choice of Location," *Journal of Industrial Economics,* **24** (1976), 313–320.

[79] Sakashita, N., "Production Function, Demand Function and the Location Theory of the Firm," *Papers of the Regional Science Association,* **20** (1967), 109–129.

[80] Salop, S. C., "Monopolistic Competition with Outside Goods," *Bell Journal of Economics,* **10** (1979), 141–156.

[81] Samuelson, P. A., "The Monopolistic Competition Revolution," in *Monopolistic Competition Theory,* ed. by R. Kuenne. New York: J. Wiley and Sons, 1967.

[82] Schmalensee, R., "Entry Deterrence in the Ready-to-Eat Breakfast Cereal Industry," *Bell Journal of Economics,* **9** (1978), 305–327.

[83] Schuler, R. E. and B. F. Hobbs, "Spatial Price Duopoly Under Uniform Delivered Pricing," *Journal of Industrial Economics,* **31** (1982), 175–187.

[84] Selten, R., "Reexamination of the Perfectness Concept for Equilibrium Points in Extensive Games," *International Journal of Game Theory,* **4** (1975), 25–55.

[85] Shaked, A., "Existence and Computation of Mixed Strategy Nash Equilibrium for 3-Firms Location Problem," *Journal of Industrial Economics,* **31** (1982), 93–96.

[86] Shaked, A. and J. Sutton, "Natural Oligopolies," *Econometrica,* **51** (1983), 1469–1483.

[87] Shapley, L. S., "Some Topics in Two-Person Games," in *Advances in Game Theory,* ed. by M. Dresher, L. S. Shapley and A. W. Tucker. Princeton: Princeton University Press, 1964.

[88] Shilony, Y., "Mixed Pricing in Oligopoly," *Journal of Economic Theory,* **14** (1977), 373–388.

[89] Shilony, Y., "Hotelling's Competition with General Customer Distributions," *Economics Letters,* **8** (1981), 39–45.

[90] Spence, A. M., "Entry, Capacity, Investment and Oligopolistic Pricing," *Bell Journal of Economics,* **8** (1977), 534–544.

[91] Sraffa, P., "The Laws of Return Under Competitive Conditions," *Economic Journal,* **36** (1926), 535–550.

[92] Stahl, K., "Differentiated Products, Consumer Search, and Locational Oligopoly," *Journal of Industrial Economics,* **31** (1982), 97–113.

[93] Stevens, B. H. and C. P. Rydell, "Spatial Demand Theory and Monopoly Price Policy," *Papers of the Regional Science Association,* **17** (1966), 195–204.

[94] Stuart, C., "Search and the Spatial Organization of Trading," in *Studies in the Economics of Search,* ed. by S. Lippman and J. J. McCall. Amsterdam: North-Holland, 1979.

[95] Thisse, J.-F., J. E. Ward and R. E. Wendell, "Some Properties of Location Problems with Block and Round Norms," *Operations Research,* **32** (1984), 1309–1327.

[96] Varian, H. R.: "A Model of Sales," *American Economic Review,* **70** (1980), 651–659.

[97] Ward, J. E. and R. E. Wendell, "Measuring Distances via Block Norms With an Application to Facility Location Models," *Operations Research,* **33** (1985), 1074–1090.
[98] Weber, A., *Ueber den Standort der Industrien.* Tübingen: J. C. B. Mohr, 1909. English translation: *The Theory of the Location of Industries.* Chicago: Chicago University Press, 1929.
[99] Wendell, R. E. and A. P. Hurter, "Location Theory, Dominance, and Convexity," *Operations Research,* **21** (1973), 314–320.
[100] Witzgall, C., "Optimal Location of a Central Facility: Mathematical Models and Concepts," National Bureau of Standards, Report 8388, 1964.

Appendix

PROPOSITION 1 *Let* $t(s, s') = c\,|s - s'| + d(s - s')^2$.

i) *When* $0 < a < \frac{1}{6}$, *there exists no price equilibrium iff* $c/d > 16a^2/(1-2a)^2$.

ii) *When* $\frac{1}{6} \leqslant a < \frac{1}{4}$, *there exists no price equilibrium iff* $c/d > 2a/(1-4a)$.

Proof Assume that (p_1^*, p_2^*) is a price equilibrium. Three cases may arise.

In the first one, (p_1^*, p_2^*) belongs to

$$\mathscr{D}_1 = \{(p_1, p_2); \tfrac{1}{2} + a < \bar{s}(p_1, p_2) \leqslant 1\}.$$

In $\mathscr{D}_1$, we have $\bar{s}(p_1, p_2) = (p_2 - p_1 + 2ad - 2ac)/4ad$ which implies

$$P_1(p_1, p_2) = p_1 \frac{p_2 - p_1 + 2ad - 2ac}{4ad}$$

and

$$P_2(p_1, p_2) = p_2 \frac{p_1 - p_2 + 2ad + 2ac}{4ad}.$$

It is easy to show that the solutions $\hat{p}_1$ and $\hat{p}_2$ of the first-order conditions $dP_1/dp_1 = 0$ and $dP_2/dp_2 = 0$ are such that $\bar{s}(\hat{p}_1, \hat{p}_2) > \frac{1}{2} + a$. Thus, (p_1^*, p_2^*) does not belong to the interior of $\mathscr{D}_1$ and must therefore satisfy $\bar{s}(p_1^*, p_2^*) = 1$. But then, p_2^* is not the best reply of firm 2 against p_1^* since $P_2(p_1^*, p_2^*) = 0$. Consequently, no price equilibrium belongs to $\mathscr{D}_1$.

In the second case, we have (p_1^*, p_2^*) in

$$\mathscr{D}_2 = \{(p_1, p_2); 0 \leqslant \bar{s}(p_1, p_2) < \tfrac{1}{2} - a\}.$$

An argument similar to the above one, but interchanging the indices 1 and 2 shows that $\mathcal{D}_2$ contains no price equilibrium.

In the third case, (p_1^*, p_2^*) belongs to the domain

$$\mathcal{D}_3 = \{(p_1, p_2); \tfrac{1}{2} - a \leqslant \bar{s}(p_1, p_2) \leqslant \tfrac{1}{2} + a\}.$$

In $\mathcal{D}_3$, $\bar{s}(p_1, p_2)$ is given by $(p_2 - p_1 + 2ad + c)/(4ad + 2c)$ so that

$$P_1(p_1, p_2) = p_1 \frac{p_2 - p_1 + 2ad + c}{4ad + 2c}$$

and

$$P_2(p_1, p_2) = p_2 \frac{p_1 - p_2 + 2ad + c}{4ad + 2c}.$$

Some simple calculations then show that $p_1^* = p_2^* = 2ad + c$ while $P_1(p_1^*, p_2^*) = P_2(p_1^*, p_2^*) = ad + c/2$. Given p_2^*, we now study the best reply $\bar{p}_1$ of firm 1 in

$$\mathcal{A}_1 = \{p_1; 0 \leqslant s(p_1, p_2^*) < \tfrac{1}{2} - a\}$$

with

$$\bar{s}(p_1, p_2^*) = \frac{p_2^* - p_1 + 2ad + 2ac}{4ad},$$

and in

$$\mathcal{A}_2 = \{p_1; \tfrac{1}{2} + a < \bar{s}(p_1, p_2^*) \leqslant 1\}$$

with

$$\bar{s}(p_1, p_2^*) = \frac{p_2^* - p_1 + 2ad - 2ac}{4ad},$$

respectively. Consider first $\mathcal{A}_1$. The maximum of $p_1\bar{s}(p_1, p_2^*)$ over $[0, \infty[$ is reached at $\bar{p}_1 = 2ad + ac + c/2$. Then we have $\bar{s}(\bar{p}_1, p_2^*) > \frac{1}{2} - a$, from which it follows that there is no best reply against p_2^* in $\mathcal{A}_1$. Let us now come to $\mathcal{A}_2$. It is easy to check that the maximum of $p_1\bar{s}(p_1, p_2^*)$ over $[0, \infty[$ is reached at $p_1' = 2ad + c/2 - ac$. For $c/d \leqslant 4a/(1-2a)$, $\bar{s}(p_1', p_2^*) \leqslant 1$ so that $\bar{p}_1 = p_1'$. On the other hand, for $c/d > 4a/(1-2a)$, $\bar{s}(p_1', p_2^*) > 1$ which implies that p_1' is not the best reply of firm 1 against p_2^*. Actually, in this case, $\bar{p}_1$ corresponds to the price for which $1 - \bar{s}(p_1, p_2^*) = 0$, i.e., $\bar{p}_1 = c(1-2a)$.

In conclusion, we have: $\bar{p}_1 = 2ad + c/2 - ac$ for $c/d \leqslant 4a/(1-2a)$ and $\bar{p}_1 = c(1-2a)$ for $c/d > 4a/(1-2a)$. First, assume that $c/d \leqslant 4a/(1-2a)$. Clearly, $P_1(\bar{p}_1, p_2^*) > P_1(p_1^*, p_2^*)$ iff $c/d > 16a^2/(1-2a)^2$. Furthermore, $16a^2/(1-2a)^2 < 4a/(1-2a)$ when $a < \frac{1}{6}$. Accordingly, if $a < \frac{1}{6}$ and $16a^2/(1-2a)^2 < c/d \leqslant 4a/(1-2a)$, there exists no price equilibrium in $\mathscr{D}_3$. Second, suppose that $c/d > 4a/(1-2a)$. Clearly, $P_1(\bar{p}_1, p_2^*) > P_1(p_1^*, p_2^*)$ iff $c/d > 2a/(1-4a)$, assuming that $a < \frac{1}{4}$. Now, $4a/(1-2a) >$ (resp. $\leqslant$)$2a/(1-4a)$ iff $a <$ (resp. $\geqslant$)$\frac{1}{6}$. Hence, there exists no price equilibrium in $\mathscr{D}_3$ if $a < \frac{1}{6}$ and $16a^2/(1-2a)^2 < c/d$ or if $\frac{1}{6} \leqslant a < \frac{1}{4}$ and $2a/(1-4a) < c/d$.

Given the symmetry of the locations, the above-mentioned conditions are similarly derived in the case of firm 2, assuming that firm 1 charges p_1^*. Consequently, these conditions, which have been shown to be sufficient, are also necessary for $\mathscr{D}_3$ not to contain a price equilibrium. This completes the proof. □

PROPOSITION 2 *Let $t(s, s') = c\,|s - s'| + d(s - s')^2$. When $0 < a < \frac{1}{6}$, there exists no modified ZCV price-equilibrium if*

$$\frac{16a^2}{(1-2a)^2} < \frac{c}{d} < \frac{4a}{1-2a}.$$

Proof We know from the argument developed in the proof of Proposition 1 that, for $a < \frac{1}{6}$, the strategy $p_1^* = 2ad + c$ is dominated by $\bar{p}_1 = 2ad + (c/2) - ac$, with $\bar{s}(\bar{p}_1, p_2^*) < 1$ iff

$$\frac{16a^2}{(1-2a)^2} < \frac{c}{d} < \frac{4a}{1-2a}.$$

Thus, in this case, no modified ZCV price equilibrium belongs to $\mathscr{D}_3$. Furthermore, it is clear from the same proof that neither $\mathscr{D}_1$ nor $\mathscr{D}_2$ contains a modified ZCV price equilibrium since there are no local maximizers of P_1 and P_2 in those domains. □

Consider a circle with a unit circumference and denote by $\bar{\delta}$ the shortest distance between firms 1 and 2; clearly $\bar{\delta} \leqslant \frac{1}{2}$.

PROPOSITION 3 *Let $t(s, s') = c\,|s - s'|$. When $\bar{\delta} < \frac{1}{4}$ there exists no price equilibrium.*

Proof If (p_1^*, p_2^*) is a price equilibrium, then (p_1^*, p_2^*) must belong to $\mathscr{D}_1 = \{(p_1, p_2); D_1(p_1, p_2) > 0$ and $D_2(p_1, p_2) > 0\}$. In $\mathscr{D}_1$, the demand functions can be shown to be given by $D_1(p_1, p_2) = [2(p_2 - p_1) + c]/2c$ and $D_2(p_1, p_2) = [2(p_1 - p_2) + c]/2c$. Thus, applying the first-order conditions yields $p_1^* = p_2^* = c/2$, while $P_1(p_1^*, p_2^*) = P_2(p_1^*, p_2^*) = c/4$. Now, in the domain $\{p_1; D_2(p_1, p_2^*) = 0\}$, the best reply of firm 1 against p_2^* is given by $\bar{p}_1 = (c/2) - c\bar{\delta} - \varepsilon$, with $\varepsilon > 0$ arbitrarily small. The corresponding profit is $P_2(\bar{p}_1, p_2^*) = (c/2) - c\bar{\delta} - \varepsilon$. Accordingly, p_1^* is the best reply of firm 1 against p_2^* over $[0, \infty[$ iff $P_1(p_1^*, p_2^*) \geqslant P_1(\bar{p}_1, p_2^*)$, i.e., $\bar{\delta} \geqslant \frac{1}{4}$. Clearly, a similar condition obtains for firm 2.

PROPOSITION 4 *Let $t(s, s') = c\,|s - s'| + d(s - s')^2$. If firm 1 is located at $s_1 \geqslant 1$ and firm 2 at $s_2 > s_1$, then there is a unique price equilibrium given by*

$$p_1^*(s_1, s_2) = (s_2 - s_1) \cdot \frac{d(s_1 + s_2 + 2) + c}{3}$$

$$p_2^*(s_1, s_2) = (s_2 - s_1) \cdot \frac{d(4 - s_1 - s_2) - c}{3}$$

when

$$\frac{c}{d} < 4 - s_1 - s_2$$

and by

$$p_1^*(s_1, s_2) = (s_2 - s_1) \cdot [d(s_1 + s_2 - 2) + c]$$
$$p_2^*(s_1, s_2) = 0$$

when

$$\frac{c}{d} \geqslant 4 - s_1 - s_2.$$

Proof Let (p_1^*, p_2^*) be a price equilibrium. (Since the profit functions are quasi-concave, we know that such an equilibrium exists.) Assume, first, that $c/d < 4 - s_1 - s_2$ (*). Over the domain

$$\mathscr{D}_1 = \{(p_1, p_2); D_1(p_1, p_1) > 0 \quad \text{and} \quad D_2(p_1, p_2) > 0\},$$

the demand functions are given by

$$D_1(p_1, p_2) = \frac{p_2 - p_1 + c(s_2 - s_1) + d(s_2^2 - s_1^2)}{2d(s_2 - s_1)}$$

and

$$D_2(p_1, p_2) = \frac{p_1 - p_2 + 2d(s_2 - s_1) - c(s_2 - s_1) - d(s_2^2 - s_1^2)}{2d(s_2 - s_1)}.$$

If $(p_1^*, p_2^*) \in \mathscr{D}_1$, then p_1^* and p_2^* must satisfy the first-order conditions $dP_i/dp_i = 0$, $i = 1, 2$. Accordingly,

$$p_1^*(s_1, s_2) = (s_2 - s_1) \cdot \frac{d(s_1 + s_2 + 2) + c}{3}$$

and

$$p_2^*(s_1, s_2) = (s_2 - s_1) \cdot \frac{d(4 - s_1 - s_2) - c}{3}.$$

Some simple calculations show that $p_2^* > 0$ and $(p_1^*, p_2^*) \in \mathscr{D}_1$ iff (*) holds. For the above prices to be the equilibrium prices under (*), it remains to prove that $P_i(p_i^*, p_j^*) > P_i(\bar{p}_i, p_j^*)$, where $\bar{p}_i$ is a best reply of firm i against p_j^* in the domain

$$\mathscr{A}_i = \{p_i; D_j(p_i, p_j^*) = 0\}.$$

Let $i = 1$. Then $\bar{p}_1 = \frac{2}{3}(s_2 - s_1) \cdot [d(s_1 + s_2 - 1) + c]$. Using (*), some simple calculations yield that $P_1(p_1^*, p_2^*) > P_1(\bar{p}_1, p_2^*)$. Furthermore, $\mathscr{A}_2$ is empty because of (*).

Assume now that $c/d \geqslant 4 - s_1 - s_2$ (**). Then, it follows from the above that (p_1^*, p_2^*) must belong to

$$\mathscr{D}_2 = \{(p_1, p_2); D_1(p_1, p_2) = 1\}.$$

That p_2^* must be equal to zero follows from the fact that, otherwise, firm 2 could decrease its price and capture a strictly positive market share. Let

$$p_1^L = (s_2 - s_1)[d(s_1 + s_2 - 2) + c]$$

be the solution to $D_1(p_1, 0) = 1$. Clearly, p_1^L dominates any price $p_1 < p_1^L$. Moreover, it can be shown that $dP_1/dp_1 < 0$ for $p_1 > p_1^L$ if (**) is satisfied. Consequently, we have $p_1^* = p_1^L$. □

A few terms that are used in the proof of Proposition 5 need to be defined.

i) Two outlets of firms 1 and 2 are said to be paired if they are located at the same point.

ii) We say that two outlets of firm i (located at s_{i1} and s_{i2}) sandwich one outlet of firm j (located at s_j) when $|s_{i1} - s_{i2}| = 2\varepsilon$ and $|s_{i1} - s_j| = |s_{i2} - s_j| = \varepsilon$ with $\varepsilon > 0$ arbitrarily small.

PROPOSITION 5 *Let $1/2f$ be an integer. Then, there exists a unique outlet selection equilibrium given by $s^*_{1i} = s^*_{2i} = (2i - 1)/2n^*$, $i = 1, \ldots, n^*$, with $n^* = 1/2f$.*

Proof First, let us show that $(\mathbf{s}^*_1, \mathbf{s}^*_2)$ is an outlet selection equilibrium. We have $P_i(\mathbf{s}^*_1, \mathbf{s}^*_2) = \frac{1}{2} - n^*f = 0$. Let, say, $\mathbf{s}_1 \neq \mathbf{s}^*_1$. If $m > n$, then firm 1 cannot do better than sandwiching $m - n^*$ outlets of firm 2 and pairing its remaining outlets with $2n^* - m$ outlets of firm 2. In this case, $D_1(\mathbf{s}_1, \mathbf{s}^*_2) < m/2n^*$ so that $P_1(\mathbf{s}_1, \mathbf{s}^*_2) < 0$. If $m \leqslant n$, then firm 1 cannot do better than pairing its outlets with m outlets of firm 2. Hence, $D_1(\mathbf{s}_1, \mathbf{s}^*_2) \leqslant m/2n^*$ so that $P_1(\mathbf{s}_1, \mathbf{s}^*_2) \leqslant 0$. Accordingly, $P_1(\mathbf{s}^*_1, \mathbf{s}^*_2) \geqslant P_1(\mathbf{s}_1, \mathbf{s}^*_2)$ for any $\mathbf{s}_1 \neq \mathbf{s}^*_1$.

Second, let $(\hat{\mathbf{s}}_1, \hat{\mathbf{s}}_2)$ be an outlet selection equilibrium. Assume that $m > n$ (the case $n > m$ is perfectly similar). Then, some outlets of firm 2 must be sandwiched by outlets of firm 1, otherwise firm 1 could increase its demand. But then, those outlets of firm 2 have an infinitesimal market area, so that firm 2 could increase its profits by deleting them, a contradiction. This means that no equilibrium exists with $m \neq n$. Suppose now that $m = n$. Then, the first outlets of firms 1 and 2 must be paired at $1/2n$. By induction, it can be shown that $\hat{s}_{1i} = \hat{s}_{2i} = (2i - 1)/2n$. It remains to show that $n = n^*$. If $n > n^*$, then $P_i(\hat{\mathbf{s}}_1, \hat{\mathbf{s}}_2) = \frac{1}{2} - nf < 0$, a contradiction. If $n < n^*$, then $P_i(\hat{\mathbf{s}}_1, \hat{\mathbf{s}}_2) < \frac{1}{2} + (\frac{1}{2} - \varepsilon) - (n + 1)f$ which means that firm i could increase its profits by adding a new outlet and by sandwiching one outlet of its competitor, a contradiction. Consequently, we have $n = n^*$. □

Urban Land Use Theory

MASAHISA FUJITA[1]

University of Pennsylvania, Philadelphia, PA, USA

0. INTRODUCTION

The modern theory of urban land use is essentially a revival of von Thünen's theory [147] of agricultural land use. Despite its significance as a monumental contribution to scientific thought, von Thünen's theory has languished without attracting the widespread attention of economists for over a century.[2] During that time human settlements grew extensively and outpaced the traditional guidance of urban design. It was the outburst of urban problems since the late 1950's that manifested the urgent need for a systematic theory of urban space and brought back the attention of location theorists and economists to von Thünen's theory.

Isard [68, Ch. 8] first suggested that von Thünen's theory could be extended in an urban context. Following the pioneering works of Beckmann [19] and Wingo [154], Alonso [2] succeeded in generalizing the concept of *bid rent curves,* the essence of von Thünen's theory, in an urban context. Since then, the subject has advanced considerably, inspiring a great deal of theoretical and empirical work. These efforts have culminated in Muth [93], Mills [81, 82],

[1] The author gratefully acknowledges the valuable suggestions and comments offered by Richard Arnott and Yoshitsugu Kanemoto. He is also grateful to Hiroyuki Koide, Elizabeth Titus and Steffen Ziss for their patient work in editing this article. This material is based upon work supported by NSF grants SOC 78-12888, SES 80-14527 and SES 85-02886 which are gratefully acknowledged.

[2] For an excellent appraisal of Thünen's achievements from the viewpoint of modern economic theory, see Samuelson [123]. Samuelson states that "Thünen belongs to the Pantheon with Leon Walras, John Stuart Mill, and Adam Smith" (Samuelson [123], p. 1482]).

Henderson [61], Kanemoto [70] and Miyao [89], to name a few. Today, the theory of urban land use represents one of the youngest and liveliest fields in economics and regional science. This paper examines the state of the art in the economic theory of urban land use, including both positive and normative aspects of the theory. In most western societies, land is allocated among alternative uses mainly in private markets, with more or less public regulations. Many studies have revealed that strong regularities exist in the spatial structure of different urban areas. Positive theory provides explanations for these regularities. The existence of regularities, however, does not necessarily imply that the spatial structure of a city is a desirable one. It is the task of normative theory to identify the efficient spatial structure, and to suggest the means for achieving it.

The theory of urban land use is an especially appealing topic of study because traditional economic theory can not be readily applied. For although traditional theory aptly describes the competitive land market typical of most Western societies, it was designed to deal with spaceless problems. With land use, such considerations as spatial contiguity, externalities and durability of buildings become of vital importance. First, empirically one generally finds that households, as well as many firms and government agencies, choose one and only one location. In the terminology of traditional economic theory, this implies that there is a strong nonconvexity in consumption and production sets. Secondly, since the essence of cities is the presence of many people and firms in a limited area, externalities are a common feature. Public services, noise, pollution, and traffic congestion all involve externalities. Moreover, the necessity of non-price interactions such as information exchanges through face-to-face communication is one of the major reasons for people and firms to locate in a city. Finally, buildings and other urban infrastructure are among the most durable objects we make, and this limits the usefulness of the static theory. Because many spatial phenomena such as urban sprawl and renewal can be satisfactorily treated only in a dynamic framework, we eventually need to combine urban land use theory with capital theory. Clearly, the city is a fertile ground for economic study.

The first three sections of this paper concern the static theory of urban land use. Section 1 presents the basic theory of residential land use within the context of the monocentric city in the absence of externalities. This section also introduces the main concepts and approach in the urban land use theory. Section 2 extends the basic theory by introducing public goods, neighborhood externalities and transport congestion. These two sections assume the location of firms is exogenously given, and are only concerned with the determination of household location. In Section 3, those models which simultaneously determine the location of households and firms are discussed. Finally, in Section 4, dynamic models of urban land use which explicitly consider the durability and adjustment cost of urban infrastructure are presented.

This paper aims not only to review the past works but also to present them in a coherent unified framework. For this purpose, we adopt the *bid rent function approach* which was introduced into an agricultural land use model by von Thünen [147], and later extended into the urban context by Alonso [2]. This approach is essentially the same as the indirect utility function approach which was introduced into an urban land use model by Solow [136]. A bid rent function transforms indifference curves in commodity space into indifference curves in urban space, i.e., bid rent curves. It is with these indifference curves defined in urban space that we will be able to graphically analyze the locational choice of the household (or firm). Moreover, since bid rent curves are stated as a pecuniary bid per unit of land, they are comparable among different land users. We will therefore be able to analyze competition for land among different agents, again, graphically in urban space. Furthermore, this bid rent function approach will enable us to conduct global analysis in contrast to the traditional local analysis via differential calculus. The same approach will be used in dynamic models of Section 4 by replacing the bid rent function with the bid (asset) price function of land.

1. BASIC THEORY OF RESIDENTIAL LAND USE

The simple models developed in this section serve as bases or building blocks for developing more complex models in the later sections. The models we will be dealing with are based on the following set of simplifying assumptions.

The city is monocentric. That is, it has a single, prespecified center of fixed size called the central business district (CBD). All job opportunities are located in the CBD. The transport system in the city is radial and dense in every direction; it is also free of congestion. The only travel is commuting of workers between residences and work places in the CBD. Travel within the CBD is ignored. The city is on a featureless plane, where all land parcels are identical, ready for residential use without further improvement. No local public goods or bads are in evidence, nor are there any neighborhood externalities.

The only spatial characteristic of each location in the city that matters to households is its distance from the CBD. Thus, the urban space can be treated as if it were one-dimensional. This is important since it greatly simplifies the analysis.

The market models presented in this section are largely due to Alonso [2] and Muth [93], the optimal models to Herbert and Stevens [62] and works by new urban economists in the early 1970s.[3]

1.1. Locational choice of the household

Imagine a household which arrives in a city and wishes to select a residence. As is typical in the economic analysis of consumer behavior, we assume that the household will maximize its utility subject to a budget constraint. We specify the utility function, $U(z, s)$, where z represents the amount of composite consumer good, and s the consumption of land, or the *lot size* of the house. The composite good is chosen as the numeraire, so its price is one. The household earns a fixed income Y per unit time which is spent on the composite good, land, and transportation. If the household is located at distance r from the CBD the budget constraint is given by $z + R(r)s = Y - T(r)$, where $R(r)$ is the unit land rent at r, and $T(r)$ the transport cost at r. Thus, we can express the residential

[3] In the early 1970's, numerous papers were written making use of optimal control or programming theory to analyze optimum or market equilibrium land use. This new approach was dubbed the "new urban economics" by Mills and Mackinnon [85]. Some of early important contributions are Dixit [33], Mills [81], Mirrlees [87], Oron, Pines and Sheshinski [102], Solow and Vickrey [137] and Solow [134, 135, 136]. For survey of the new urban economics, see Mills and MacKinnon [85], Anas and Dendrinos [8], and Richardson [112].

choice of the household as

$$\max_{r,z,s} U(z, s), \quad \text{subject to } z + R(r)s = Y - T(r), \tag{1.1}$$

which is called the *basic model* of residential choice. This model is convenient in illustrating the basic mechanism of trade-off between accessibility and space in residential choice. In the subsequent analysis, we always assume that:

Assumption 1.1 (*well behaved utility function*) The utility function is differentiable, strictly quasi-concave and strictly increasing, and indifference curves do not cut the axes.

Assumption 1.2 (*increasing transport cost*) Marginal transport cost $T'(r) \equiv dT(r)/dr$ is always positive, $T(\infty) = \infty$ and $T(0) < Y$.

Assumption 1.3 (*normality of land*) Income effect on the ordinary demand for land is positive.

By directly solving the optimization problem implied by the basic model (1.1), we could ascertain, in a straightforward manner, the household's residential decision. But there is another approach, conceptually much richer, which leads to a desirable elaboration of theory. This approach, which mimics the von Thünen model of agricultural land use, requires the introduction of a concept called bid rent: *Bid rent* $\Psi(r, u)$ *is the maximum rent per unit land the household would be able to pay for residing at distance r while enjoying a fixed level of utility u.* Formally, bid rent may be defined as

$$\Psi(r, u) = \max_{z,s} \left\{ \frac{Y - T(r) - z}{s} \mid U(z, s) = u \right\}. \tag{1.2}$$

Or, we may first solve the utility constraint, $U(z, s) = u$, for z, and obtain the equation of the indifference curve as $z = Z(s, u)$. Then, the bid rent function can be redefined as

$$\Psi(r, u) = \max_{s} \frac{Y - T(r) - Z(s, u)}{s}, \tag{1.3}$$

which is an unconstrained maximization problem.[4] When we solve

[4] Note that because of Assumption 1.1, whenever it exists, the optimal s for the maximization problem of (1.3) is positive. When there is no solution for this maximization problem, we define $\Psi(r, u) = 0$ and $S(r, u) = \infty$.

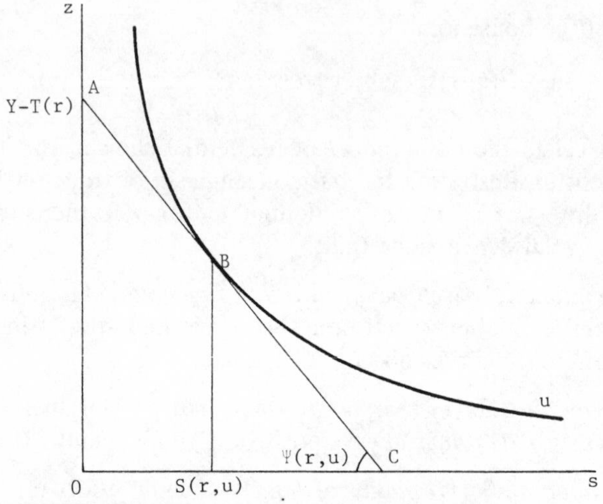

FIGURE 1.1 Bid rent $\Psi(r, u)$ and bid-max lot size $S(r, u)$.

the maximization problem of (1.2) or (1.3), we obtain the optimal lot size, $S(r, u)$, which is called the *bid-max lot size.* In order to emphasize that the basic parameters of bid rent $\Psi(r, u)$ and bid-max lot size $S(r, u)$ are net income $Y - T(r)$ and utility level u, we often express them as

$$\Psi(r, u) \equiv \psi(Y - T(r), u), \qquad S(r, u) = s(Y - T(r), u). \quad (1.4)$$

Graphically, as depicted in Figure 1.1, bid rent $\Psi(r, u)$ is given by the slope of the budget line at distance r which is just tangent to indifference curve u;[5] bid-max lot size $S(r, u)$ is determined from the tangency point B.

EXAMPLE 1 In the case of a log-linear utility function,

$$U(z, s) = \alpha \log z + \beta \log s, \quad \text{where} \quad \alpha > 0, \beta > 0, \alpha + \beta = 1,$$

we have $Z(s, u) = s^{-\beta/\alpha} e^{u/\alpha}$. Solving the maximization problem of (1.3) with this utility function, we have

$$\Psi(r, u) = \alpha^{\alpha/\beta} \beta (Y - T(r))^{1/\beta} e^{-u/\beta}, \quad (1.5)$$

$$S(r, u) = \beta(Y - T(r))/\Psi(r, u) = \alpha^{-\alpha/\beta}(Y - T(r))^{-\alpha/\beta} e^{u/\beta}. \quad (1.6)$$

[5] More precisely, if we denote the angle ACO by θ, then $\Psi(r, u) = \tan \theta$. But, for simplicity, we use this graphical expression throughout the paper.

In order to relate bid rent $\Psi(r, u)$ and bid-max lot size $S(r, u)$ to familiar microeconomic notions, we may appeal, once again, to Figure 1.1. In Figure 1.1 we may interprete that indifference curve u is tangent to budget line AC from above at point B. This implies the following: if we define the *ordinary demand*, $\hat{s}(I, R)$, for land from the solution of the next utility maximization problem,

$$\max_{z,s} U(z, s), \text{ subject to } z + Rs = I,$$

then it holds identically that

$$S(r, u) \equiv \hat{s}(Y - T(r), \Psi(r, u)). \tag{1.7}$$

Or, we may interpret Figure 1.1 as budget line AC being tangent to indifference curve u from below at B. This implies that: if we define the *compensated demand*, $\tilde{s}(R, u)$, for land from the solution of the next expenditure minimization problem,

$$\min_{z,s} z + Rs, \quad \text{subject to } U(z, s) = u,$$

then it holds identically that

$$S(r, u) \equiv \tilde{s}(\Psi(r, u), u). \tag{1.8}$$

Identities (1.7) and (1.8) turn out to be very useful in deriving qualitative results.[6]

Next, we examine important properties of bid rent and bid-max lot size functions. Through an application of the envelope theorem to (1.3), we have

$$\frac{\partial \Psi(r, u)}{\partial r} = -\frac{T'(r)}{S(r, u)} < 0. \tag{1.9}$$

Hence, from identity (1.8),

$$\frac{\partial S(r, u)}{\partial r} = \frac{\partial \tilde{s}}{\partial R} \frac{\partial \Psi(r, u)}{\partial r} > 0, \tag{1.10}$$

[6] We can also derive the following identities. Define the *indirect utility function*, $V(I, R) = \max_{z,s} \{U(z, s) \mid z + Rs = I\}$; define the expenditure function, $E(R, u) = \min_{z,s} \{z + Rs \mid U(z, s) = u\}$. And, put $I(r) = Y - T(r)$. Then, we can see from Figure 1.1 that $V(I(r), \psi(I(r), u)) \equiv u$ and $E(\psi(I(r), u), u) \equiv I(r)$. That is, ψ and V are inverse to each other as are ψ and E. In fact, in Solow [136], bid rent function ψ has been defined by solving the relation, $V(I, R) = u$, for R.

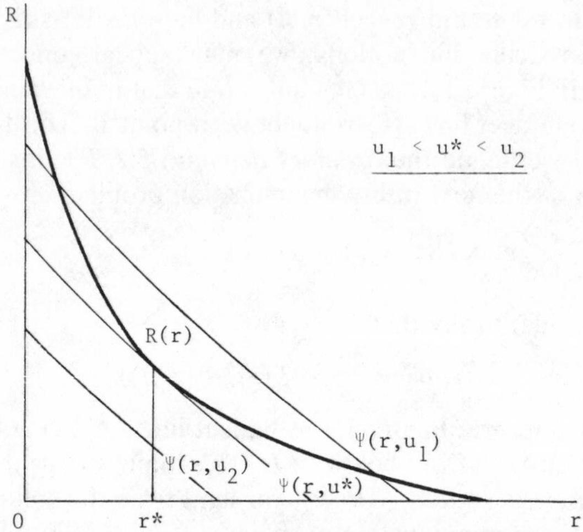

FIGURE 1.2 Determination of the equilibrium location.

which is positive since $\partial \tilde{s}/\partial R$ is always negative. Similarly,

$$\frac{\partial \Psi(r, u)}{\partial u} = -\frac{1}{S(r, u)} \frac{\partial Z(s, u)}{\partial u} < 0, \tag{1.11}$$

which is negative since $\partial Z(s, u)/\partial u$ is positive from Assumption 1.1. So, from identify (1.7),

$$\frac{\partial S(r, u)}{\partial u} = \frac{\partial \hat{s}}{\partial R} \frac{\partial \Psi(r, u)}{\partial u} > 0, \tag{1.12}$$

which is positive since $\partial \hat{s}/\partial R$ is negative from Assumption 1.3. Therefore, we can conclude that *bid rent* $\Psi(r, u)$ *is decreasing in both r and u, and bid-max lot size* $S(r, u)$ *is increasing in both r and u.*[7]

We are now ready to explain how the equilibrium (or optimal)

[7] If we further assume that transport cost is linear or concave in r (i.e., $T''(r) \leqq 0$), then from (1.9) and (1.10),

$$\frac{\partial^2 \Psi(r, u)}{\partial r^2} = -\frac{T''(r)}{S(r, u)} + \frac{T'(r)}{S(r, u)} \frac{\partial S(r, u)}{\partial r} > 0,$$

which means bid rent curves are strictly convex in r.

location of the household is determined, given the basic model (1.1) and the land rent curve $R(r)$ of the city. In Figure 1.2, the market rent curve $R(r)$ is depicted; and a set of bid rent curves are superimposed on it. It is evident from the figure that *the equilibrium location of the household is distance r^* at which a bid rent curve $\Psi(r, u^*)$ is tangent to the market rent curve $R(r)$ from below.* That is, when the household decides to locate somewhere in the city, it is obliged to pay the market land rent. At the same time, the household will maximize its utility. Since the utility of bid rent curves increases towards the origin, the highest utility will be achieved at a location at which a bid rent curve is tangent to the market rent curve from below. This conclusion can be formally stated as follows:

Rule 1.1. Given the market rent curve $R(r)$, u^* is the equilibrium utility level of the household, and r^* is an optimal location if and only if $R(r^*) = \Psi(r^*, u^*)$ and $R(r) \geqq \Psi(r, u^*)$ for all r.

Given that curves $R(r)$ and $\Psi(r, u^*)$ are smooth at r^*, the above rule implies

$$\frac{\partial \Psi(r^*, u^*)}{\partial r} = R'(r^*). \tag{1.13}$$

Thus, recalling equation (1.9), we have

$$T'(r^*) = -R'(r^*)S(r^*, u^*) \tag{1.14}$$

This result, called *Muth's Condition,* asserts that at the equilibrium location the marginal transport cost $T'(r)$ equals the marginal land cost saving, $-R'(r)\, S(r, u^*)$.

Next, we can study the relative locations of different households having different bid rent functions. A general rule for ordering equilibrium locations of different households with respect to the distance from the CBD is as follows:

Rule 1.2. If the equilibrium bid rent curve $\Psi_i(r, u_i^*)$ of household i and equilibrium bid rent curve $\Psi_j(r, u_j^*)$ of household j intersect only once and if $\Psi_i(r, u_i^*)$ is steeper than $\Psi_j(r, u_j^*)$ at the intersection, then the equilibrium location of household i is closer to the CBD than that of household j.

This result is depicted in Figure 1.3. In order to apply this rule, however, we must know beforehand which equilibrium curve is

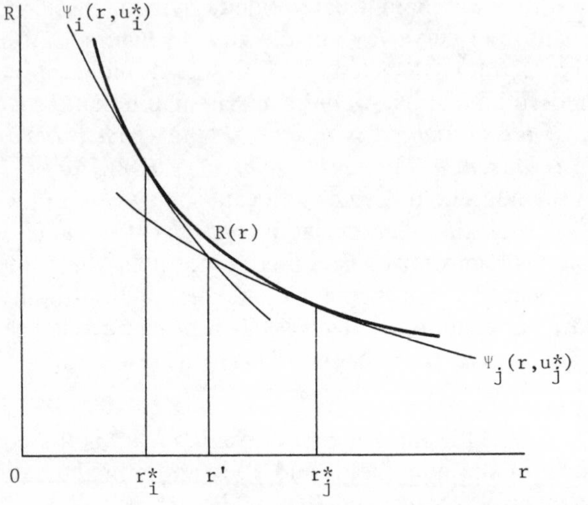

FIGURE 1.3 Ordering of equilibrium locations.

steeper at the intersection. In general, this information is difficult to obtain *a priori.* Matters can be greatly simplified, however, if we are able to determine the *relative steepness of bid rent functions.* We say that bid rent function Ψ_i is steeper than bid rent function Ψ_j if and only if *at every intersection of each pair of bid rent curves,* apiece for household i and j, the former is always steeper than the latter.[8] Figure 1.4 gives an illustration. From Rule 1.2 combined with this definition, we can state:

Rule 1.3. If the bid rent function of household i is steeper than that of household j, then the equilibrium location of household i is closer to the CBD than that of household j.

Although the applicability of this rule is limited, it becomes very useful in comparative static analysis, where the effects of difference in model parameter values are examined. In fact, almost always when a definite conclusion can be obtained from a comparative static analysis of household location, the relative steepness of bid

[8] Stated formally, function Ψ_i is steeper than function Ψ_j if and only if the following condition is met: whenever $\Psi_i(r, u_i) = \Psi_j(r, u_j) > 0$, then either $-\partial\Psi_i(r, u_i)/\partial r > -\partial\Psi_j(r, u_j)/\partial r$, or $\partial\Psi_i(r, u_i)/\partial r = \partial\Psi_j(r, u_j)/\partial r$ and $\partial^2\Psi_i(r, u_i)/\partial r^2 < \partial^2\Psi_j(r, u_j)/\partial r^2$.

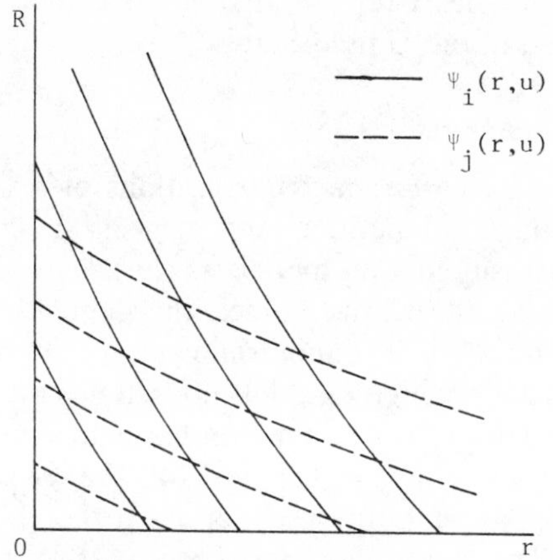

FIGURE 1.4 Relative steepness of bid rent functions.

rent functions (determined by parameter values) can be ascertained. An important example is the effect of income level on household location.[9]

In the context of basic model (1.1), let us arbitrarily specify two income levels, $Y_1 < Y_2$. It is assumed that both households possess the same utility function and face the same transport cost function. Denote by $\Psi_i(r, u)$ and $S_i(r, u)$ the bid rent and bid-max lot size of the household with income Y_i $(i = 1, 2)$. Let us arbitrarily take a pair of bid rent curves, $\Psi_1(r, u_1)$ and $\Psi_2(r, u_2)$, and suppose that they intersect at some distance $r' : \Psi_1(r', u_1) = \Psi_2(r', u_2) \equiv \bar{R}$. Recall indentity (1.7). Since $Y_1 - T(r') < Y_2 - T(r')$, from the normality of land,

$$S_1(r', u_1) = \hat{s}(Y_1 - T(r'), \bar{R}) < \hat{s}(Y_2 - T(r'), \bar{R}) = S_2(r', u_2).$$

Thus, from (1.9), $-\partial\Psi_1(r', u_1)/\partial r > -\partial\Psi_2(r', u_2)/r$. Since we have arbitrarily chosen two bid rent curves, this result means that function Ψ_1 is steeper than Ψ_2. Thus, from Rule 1.3, we may state the following proposition: *Other aspects being equal, higher income households locate further from the CBD than lower income house-*

[9] For other examples, see Sections 1.4.A and 1.4.B.

holds. This result has been often used to explain the residential pattern observed in the United States.[10]

1.2. Equilibrium land use

We next examine equilibrium configurations of residential land as determined through competitive land markets. Following Wheaton [148], we may classify market models as *closed city models* and *open city models*. In the closed city model, the population of the city is exogenously fixed, while equilibrium utility is determined endogenously. In the open city model, households are assumed to be able to move costlessly across the city boundary; hence, the utility of residents equals that of the rest of the economy which is exogeneously fixed, while the population of the city is determined endogenously. The closed city model is a useful conceptual device when analyzing urban land use in large cities or "average cities" of developed countries. On the other hand, the open city model better describes urban conditions in developing countries which have surplus labor in rural areas. In the latter case, rural life often establishes the base utility level of the economy. The emphasis here will be on the closed city model, because it is more fundamental from a theoretical point of view. In both models, we must also specify the form of land ownership. Two popular specifications are the *absentee ownership model,* in which land is owned by absentee landlords, and the *public ownership model,* in which the revenue from land is equally shared among city residents.

This subsection will examine the four cases in turn. For simplicity, let us assume that all households in the economy are identical.[11] We denote by $L(r)$ the amount of land available for

---

[10] Recall, however, that this result depends critically on the assumption that transport costs are independent of income. A completely reversed spatial pattern can be observed in many European, Latin American, and Asian cities. In the United States as well, luxury apartments and townhouses are often found near the urban center. See Alonso [2, Ch. 6], Muth [93] and Wheaton [150] for empirical studies on household location. These observations suggest that factors other than income including value of time, family structure, externalities, and dynamic factors also affect residential choices, and spatial patterns. These factors will be introduced one by one in the rest of the paper.

[11] The equilibrium pattern of residential land with multiple income classes was studied by Muth [93], Solow [136], Hartwick, Schweizer and Varaiya [56], Ando [10] and Fujita [45, 46]. Beckmann [20], with some corrections by Montesano [91], considered the case of a continuous, Pareto income distribution. See Miyao [88] for a dynamic stability analysis of boundaries between different income classes.

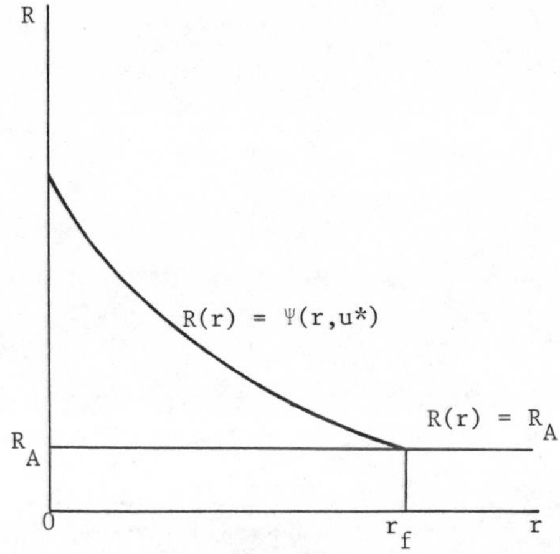

FIGURE 1.5 Equilibrium spatial configuration.

residential use at each distance r, which is assumed to be positive for all $r > 0$. It is also assumed that land not occupied by households is used for agriculture, yielding a constant rent R_A.

Case 1 is the closed city model under absentee land ownership. There are M identical households in the city. We assume that households behave according to the basic model (1.1), and that the household income Y is given exogenously.[12] Then, the bid rent function $\Psi(r, u)$ and bid-max lot size function $S(r, u)$ can be derived as explained before. Let u^*, $R(r)$, r_f and $n(r)$ be the utility level, land rent curve, urban fringe distance, and household distribution (i.e., the number of households between r and $r + dr$ equals $n(r)dr$) in equilibrium, respectively. From Rule 1.1, the equilibrium land rent $R(r)$ must be equal to the equilibrium bid rent $\Psi(r, u^*)$ everywhere in the residential area. Thus, we have

$$R(r) = \Psi(r, u^*) \quad \text{for} \quad r \leqq r_f, \qquad = R_A \quad \text{for} \quad r \geqq r_f. \tag{1.15}$$

This relationship is described in Figure 1.5. The household distribution is given by

$$n(r) = L(r)/S(r, u^*) \quad \text{for} \quad r < r_f. \tag{1.16}$$

[12] For endogenous determination of wage income Y, see, for example, Solow [136], Henderson [61] and Kanemoto [70].

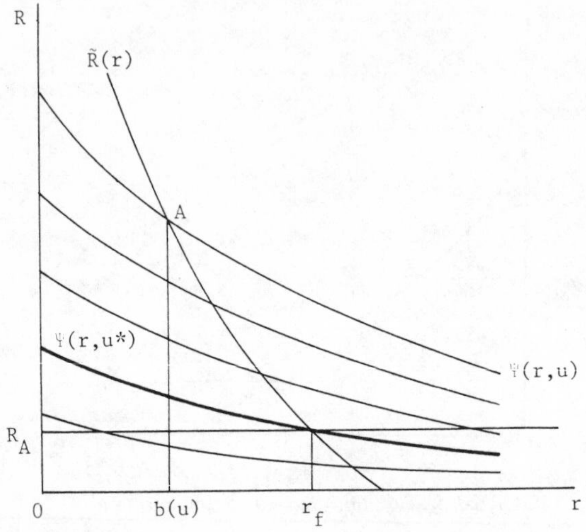

FIGURE 1.6 Boundary rent curve $\tilde{R}(r)$ for equilibrium land use.

Hence, the population constraint is given by

$$\int_0^{r_f} \frac{L(r)}{S(r, u^*)} dr = M. \tag{1.17}$$

Two unknowns u^* and r_f can be determined from (1.17) and the next boundary rent condition,

$$\Psi(r_f, u^*) = R_A. \tag{1.18}$$

The existence and uniqueness of the equilibrium solution can be demonstrated by using the concept of the *boundary rent curve.*[13] Under each value of u, solve the next equation for b,

$$\int_0^{b} \frac{L(r)}{S(r, u)} dr = M, \tag{1.19}$$

and obtain the *outer boundary function,* $b(u)$, of residential area. For each given u, $b(u)$ marks a distance on the corresponding bid rent curve $\Psi(r, u)$, such as point A in Figure 1.6. By changing u, we

[13] A similar concept was introduced by Kanemoto [70, Ch. 6], in his stability analysis of mixed cities. Fujita [45, 46] generalized it to the case of variable lot size, and used it for the study of the existence and uniqueness of the solution.

can obtain a curve, $\tilde{R}(r)$, called the *boundary rent curve,* as depicted in Figure 1.6. If the inverse of $r = b(u)$ is denoted by $U(r)$, the boundary rent curve may be defined as

$$\tilde{R}(r) \equiv \Psi(r, U(r)). \tag{1.20}$$

By definition, $\tilde{R}(r)$ represents the market land rent at r when the residential boundary occurs at r. Suppose we have obtained the curve $\tilde{R}(r)$. Then, as depicted in Figure 1.6, the equilibrium fringe distance r_f is given by the point where $\tilde{R}(r_f) = R_A$. Then, equilibrium utility level u^* can be obtained from the relationship, $\Psi(r_f, u^*) = R_A$. Under Assumptions 1.1 to 1.3, it is not difficult to show that the boundary rent curve $\tilde{R}(r)$ is continuously decreasing in r, becomes zero as r approaches a certain finite distance, and becomes infinitely high as r approaches zero (see Fujita [45]). Therefore, as demonstrated in Figure 1.6, we can conclude that *there exists a unique equilibrium for the closed city model under absentee land ownership.*[14]

For case 2, the open city model under absentee land ownership, the equilibrium is trivially simple to obtain. If the national utility level is given by a constant u^*, then the urban fringe distance r_f can be obtained from relation (1.18). The equilibrium population M can be determined by (1.17), and the land rent curve $R(r)$ is given by (1.15).

Next, for Case 3 let us consider the closed city model under public land ownership, which was introduced by Solow [136]. The city residents form a government, which rents the land for the city from rural landlords at agricultural rent R_A. The city government in turn, subleases the land to city residents at the competitively determined rent, $R(r)$, at each location. Define the *total differential rent* (TDR) from the city by

$$\text{TDR} = \int_0^{r_f} (R(r) - R_A)L(t)\, dr. \tag{1.21}$$

Assume that there are M identical households in the city. Then, the income of each household is its wage income (or non-rent income)

[14] See Fujita [45, 46] for the extension of the boundary rent curve approach for the proof of existence and uniqueness of equilibrium in the case of multiple household types.

Y plus a share of land rent, TDR/M. Thus, the residential choice behavior of each household can be formulated as follows:

$$\max_{r,z,s} U(z, s), \text{ subject to } z + R(r)s = Y + (\text{TDR/M}) - T(r). \tag{1.22}$$

Since TDR is yet unknown, we cannot apply the boundary rent curve approach in order to examine the existence and uniqueness of the equilibrium. We will prove them in the next subsection via the existence and uniqueness of the optimal land use.

Case 4 is the open city model under public land ownership. As in Case 3, each resident of the city gets a wage income Y plus a share of land rent, TDR/M. The residential choice behavior of each household can be described, as before, by (1.22). However, both TDR and M are now unknown; the utility of residents is fixed at the national utility level u. We will prove the existence and uniqueness of the equilibrium in the next subsection.

Comparative statics were studied by Wheaton [148] in the case of single household type; and by Wheaton [149], Hartwick, Schweizer and Varaiya [56], and Arnott, MacKinnon and Wheaton [14] in the case of multiple income classes. Take the case of single household type in the context of the closed city model under absentee land ownership, where the residential choice of each household is represented by the basic model (1.1). Assume for simplicity that $T(r) = ar$. Then, four parameters are agricultural rent R_A, population M, marginal transport cost a, and income Y. Under Assumptions 1.1, 1.2 and 1.3, either by traditional differential calculus (Wheaton [148], Hartwick, *et al.* [56]), or by the boundary rent curve approach (Fujita [46]), we can obtain the following results:

$$\frac{du^*}{dR_A} < 0, \quad \frac{dr_f}{dR_A} < 0, \quad \frac{dR(r)}{dR_A} > 0 \text{ and } \frac{dS(r, u^*)}{dR_A} < 0 \text{ for all } r, \tag{1.23}$$

$$\frac{du^*}{dN} < 0, \quad \frac{dr_f}{dN} > 0, \quad \frac{dR(r)}{dN} > 0 \text{ and } \frac{dS(r, u^*)}{dN} < 0 \text{ for all } r, \tag{1.24}$$

$$\frac{du^*}{da} < 0, \quad \frac{dr_f}{da} < 0, \quad \frac{dR(0)}{da} > 0, \quad \frac{dS(0, u^*)}{da} < 0, \tag{1.25}$$

$$\frac{du^*}{dY} > 0, \quad \frac{dr_f}{dY} > 0, \quad \frac{dS(0, u^*)}{dY} > 0. \tag{1.26}$$

These results are hardly surprising. On the other hand, the effect of income change on the land rent at the city center depends crucially on the distribution of land, and we can show that:

$$\frac{dR(0)}{dY} \gtreqless 0 \quad \text{as} \quad L'(r) \lesseqgtr 0 \quad \text{for all } r. \tag{1.27}$$

Since $L'(r)$ is positive in most cities, from (1.25) and (1.27) we can conclude that both a decrease in marginal transport cost and an increase in income will make the land rent curve flatter (i.e., $dR(0) < 0$ and $dr_f > 0$), and will also make the population density curve flatter (i.e., $dS(0, u^*) > 0$ and $dr_f > 0$). This result describes the suburbanization trend observed in the United States and other developed countries (at least, prior to the oil shocks at the beginning of 1970s).[15]

Finally, the following relationship between TDR (total differential rent) and TTC (total transport cost), which was obtained by Arnott [12], is noteworthy because of its simplicity and generality. In the context of the basic model of (1.1), suppose that transport cost function is linear (i.e., $T(r) = ar$), and that distribution of land is given by $L(r) = \theta r^\lambda$ where $\lambda > -1$. Then, for all four cases, a simple calculation using relation (1.14) yields that[16]

$$\text{TDR} = \text{TTC}/(\lambda + 1). \tag{1.28}$$

Hence, in the case of linear city ($\lambda = 0$), TDR = TTC; and in the case of a circular or fan-shaped city ($\lambda = 1$), TDR = TTC/2. It is not difficult to see from the derivation that as long as *each* household

[15] For empirical studies on historical trend in population density gradients, see, for example, Mills [82, 83] and Parr and Jones [107].

[16] $\text{TTC} = \int_0^{r_f} T(r)n(r)\,dr = \int_0^{r_f} ar\theta r^\lambda / S(r, u^*)\,dr = \int_0^{r_f} -R'(r)\theta r^{\lambda+1}\,dr$ (from (1.14)) $= -R_A\theta r_f^{\lambda+1} + \int_0^{r_f} R(r)(\lambda+1)\theta r^\lambda\,dr$ (from integration by parts) $= (\lambda+1)\int_0^{r_f}(R(r) - R_A)\theta r^\lambda\,dr = (\lambda+1)\text{TDR}$. In the case of fixed lot size, this relationship was obtained by Mohring [90] from ingeneous geometric reasoning.

has a *directionally linear* transport cost function, relationship (1.28) holds for both equilibrium cities and optimal cities.[17]

1.3. Optimal land use, optimal vs. equilibrium

This subsection studies the optimal allocation of residential land and households in a city, and examines the relationship between the optimal land use and equilibrium land use; in particular, the ability of competitive market to sustain an optimal allocation of residential land and households. Exactly what optimal land use is, of course, depends on how the objective function is specified. In spaceless economics, it is common to maximize a Benthamite social welfare function, which is the sum of utilities of individual households (an unweighted sum in the case of identical households). However, this is not the most convenient approach for the land use problem, because maximization of a Benthamite welfare function results in the assignment of different utility levels to identical households depending on their locations. Such a result is referred to as *Mirrlees' inequality* or *unequal treatment of equals,* and is a unique phenomenon due to the nonconvexity introduced by space.[18] Since competitive markets treat all equals equally, it is clear that the maximization of a Benthamite welfare function is not the most direct approach for investigating the efficiency of land markets. It turns out that the so called *Herbert-Stevens model* (*HS model*) is a convenient formulation of optimization problems for land use theory. In this model, the objective is to maximize net revenue subject to a set of prespecified target utility levels for all household types. The model is designed so that its solution is always efficient

[17] That is, different households may have different linear transport cost functions of which marginal transport costs may depend on the direction from the city center. If the time cost is involved in transport, we must also include the time cost in calculating TTC. It is not difficult to see from the derivation that relation (1.28) holds both in equilibrium cities and optimal cities. Moreover, if land rent $R(r)$ is replaced with land price $P(r)$ (i.e., asset price of land), the same relationship holds between the total differential land price and the total discounted value of transport costs for dynamic cities in Section 4 (under perfect foresight without tulip-mania expectation).

[18] This phenomenon was discovered by Mirrlees [87]. For the explanation of this phenomenon and for further discussion of this subject, see Riley [114, 115], Arnott and Riley [15], Levhari, Oron and Pines [78], and Kanemoto [70, Appendix I].

(i.e., Pareto-optimal), and all efficient allocations can be obtained by simply varying the target utility levels.

As before, we have a monocentric city with M identical households.[19] The utility function of each household is given by $U(z, s)$, and transport cost by $T(r)$. Suppose we arbitrarily choose a target utility level, u, and require that all households shall attain this. Next, suppose we choose a household distribution $n(r)$, a lot size function $s(r)$, and a residential fringe distance r_f which together satisfy the following land and population constraints:

$$s(r)n(r) \leqq L(r) \quad \text{at each} \quad r \leqq r_f, \tag{1.29}$$

$$\int_0^{r_f} n(r)\, dr = M. \tag{1.30}$$

Then, the *total cost* C for achieving target utility u can be calculated as follows:

$$\begin{aligned} C &= \text{transport costs} + \text{composite good costs} + \text{opportunity land costs} \\ &= \int_0^{r_f} (T(r) + Z(s(r), u) + R_A s(r))n(r)\, dr, \end{aligned} \tag{1.31}$$

where $Z(s, u)$ is the inverse of $u = U(z, s)$ for z. The problem is to choose an allocation, $(n(r), s(r), r_f)$, that minimizes the total cost C subject to land constraint (1.29) and population constraint (1.30). This problem can be more conveniently expressed in terms of net revenue. If we assume that the per capita income of the city is given by Y, then the total income of the city is MY. Let Y be some fixed number determined independently of the residential land use pattern. The *net revenue* NR from allocation $(n(r), s(r), r_f)$ is

$$\begin{aligned} \text{NR} &= MY - C \\ &= \int_0^{r_f} (Y - T(r) - Z(s(r), u) - R_A s(r))n(r)\, dr. \end{aligned} \tag{1.32}$$

Since MY is assumed to be a constant, minimization of C is equivalent to maximization of NR, and the *Herbert-Stevens*

[19] For the similar discussion of optimal land use with multiple household types, see Ando [10] and Fujita [45, 46].

problem, HS(u), can be stated as follows:[20]

$$\max_{n(r),s(r),r_f} \mathrm{NR} = \int_0^{r_f} (Y - T(r) - Z(s(r), u) - R_A s(r))n(r)\,dr,$$

subject to constraints (1.29) and (1.30).

Considering that NR represents a benefit (or cost, if negative) for the rest of the economy, it is obvious that the solution of any HS-problem is socially efficient, and any efficient allocation under the equal utility condition is a solution of some HS-problem.

Now to state the optimality conditions for the above HS-problem, let us define the *bid rent function with income subsidy* Q as follows:

$$\Psi(r, u, Q) = \max_s \frac{Y + Q - T(r) - Z(s, u)}{s}, \tag{1.33}$$

and denote the corresponding bid-max lot size by $S(r, u, Q)$.[21] Suppose Assumptions 1.1 and 1.2 hold. Then, for an allocation $(n(r), s(r), r_f)$ to be optimal, it is necessary and sufficient that there exist multipliers $R(r)$ and $\hat{Q}$ such that:[22]

$$R(r) = \Psi(r, u, \hat{Q}) \quad \text{for} \quad r \leqq r_f, \qquad = R_A \quad \text{for} \quad r \geqq r_f, \tag{1.34}$$

$$s(r) = S(r, u, \hat{Q}) \quad \text{and} \quad n(r) = L(r)/S(r, u, \hat{Q}) \quad \text{for} \quad r < r_f, \tag{1.35}$$

$$\int_0^{r_f} L(r)/S(r, u, \hat{Q})\,dr = M. \tag{1.36}$$

$R(r)$ can be interpreted as the shadow land rent at r, and $\hat{Q}$ the shadow income subsidy (or shadow cost) per household. The concept of income subsidy will arise again in comparing the optimal

---

[20] Since this is a continuous version of the problem introduced by Herbert and Stevens [62], we call it a Herbert–Stevens problem. A continuous version was introduced by Ando [10], Yang and Fujita [157], and Fujita [46]. An HS-problem is the dual to the maximization of the common utility level subject to the resource constraint of the city (i.e., a balanced budget). In fact, the latter problem formulation, introduced by Dixit [33] and Oron, Pines and Sheshinski [102], is more popular in the literature. However, the HS problem-formulation appears to be more convenient for comparison of market allocations and optimal allocations.

[21] Similarly to (1.4), we may express $\Psi(r, u, Q) \equiv \psi(Y + Q - T(r), u)$ and $S(r, u, Q) \equiv s(Y + Q - T(r), u)$.

[22] For the derivation of the following conditions, see Ando [10] or Fujita [46].

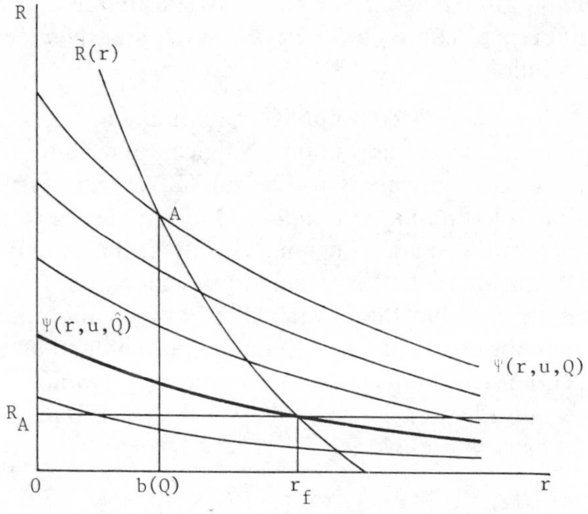

FIGURE 1.7 Boundary rent curve $\tilde{R}(r)$ for optimal land use.

land use with the equilibrium land use. Briefly, such a subsidy is one way for government to achieve the optimal solution of the $HS(u)$-problem through a competitive market.

Recalling the boundary rent curve approach, we can determine two unknowns $\hat{Q}$ and r_f. Under each value of Q, we solve the next equation for b,

$$\int_0^b \frac{L(r)}{S(r, u, Q)}\, dr = M, \tag{1.37}$$

and obtain the outer boundary function, $b(Q)$. For each given Q, $b(Q)$ marks a distance on the corresponding bid rent curve $\Psi(r, u, Q)$, such as point A in Figure 1.7. By changing Q, we can obtain the boundary rent curve, $\tilde{R}(r)$, as depicted in Figure 1.7. That is,

$$\tilde{R}(r) \equiv \Psi(r, u, Q(r)), \tag{1.38}$$

where $Q(r)$ is the inverse of $r = b(Q)$. With $\tilde{R}(r)$, then, as depicted in Figure 1.7, r_f is determined from the relation $\tilde{R}(r_f) = R_A$, and $\hat{Q}$ from $\Psi(r_f, u, \hat{Q}) = R_A$.

In addition to Assumptions 1.1, 1.2 and 1.3, suppose the following assumption is satisfied.

Assumption 1.4 z and s are perfectly substitutable. That is, on each indifference curve $u = U(z, s)$, s approaches zero as z approaches infinity.[23]

Then, as before, $\tilde{R}(r)$ is continuously decreasing in r, becomes zero as r approaches a certain finite distance, and $\lim_{r\to 0} \tilde{R}(r) = \infty$. Therefore, as demonstrated in Figure 1.7, *for any given* u *the* $HS(u)$*-problem has a unique solution.* One can also show that if we donote by $\hat{Q}(u)$ the shadow income subsidy at the solution of each HS(u)-problem, then $\hat{Q}(u)$ is strictly increasing in u.

Next, let us examine the relationship between the optimal land use and equilibrium land use. Keeping the context of the basic model (1.1), let us introduce a new parameter, Q, which represents an *income subsidy* per household. Then, the residential choice behavior of each household is:

$$\max_{r,z,s} U(z, s), \quad \text{subject to } z + R(r)s = Y + Q - T(r). \quad (1.39)$$

For the closed city model with M identical households under absentee ownership, the equilibrium conditions are:

$$R(r) = \Psi(r, u^*, Q) \quad \text{for} \quad r \leqq r_f, \quad = R_A \quad \text{for} \quad r \geqq r_f, \quad (1.40)$$

$$n(r) = L(r)/S(r, u^*, Q) \quad \text{for} \quad r < r_f, \quad (1.41)$$

$$\int_0^{r_f} L(r)/S(r, u^*, Q)\, dr = M, \quad (1.42)$$

where functions $\Psi(r, u, Q)$ and $S(r, u, Q)$ have been defined from (1.33). We may call this market problem the *Alonso-Muth problem with income subsidy* Q, and denote it by AM(Q). As in subsection 1.2, we can see that for each Q such that $Y + Q > T(0)$, the AM(Q)-problem has a unique solution. If we denote the equilibrium utility of each AM(Q)-problem by $u^*(Q)$, then $u^*(Q)$ is strictly increasing in Q.

Comparing optimality conditions (1.34)–(1.36) with equilibrium conditions (1.40)–(1.42), we can see that functions $\hat{Q}(u)$ and $u^*(Q)$ are inverse to each other: $\hat{Q}(u^*(Q)) \equiv Q$ and $u^*(\hat{Q}(u)) \equiv u$. Hence, we can conclude that for any u, the $HS(u)$-problem and $AM(\hat{Q}(u))$-

[23] If we assume that the total amount of land, $\int_0^\infty L(r)\, dr$, is infinite, then this assumption is not necessary for the existence and uniqueness of the solution.

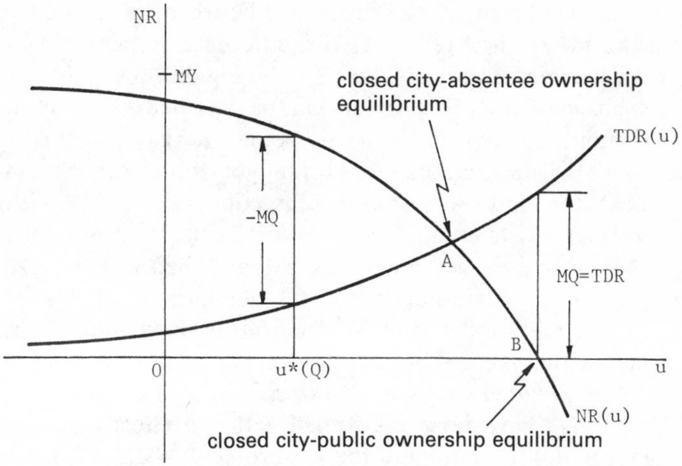

FIGURE 1.8 Relationship between optimal solutions and equilibrium solutions (M fixed).

problem have the same solution; similarly, for any Q such that $Y+Q>T(0)$, the AM(Q)-problem and HS($u^*(Q)$)-problem have the same solution. Therefore, the solution of any HS-problem can be obtained by using the AM-model and choosing an appropriate income subsidy; conversely, the solution of any AM-problem can be obtained by using the HS-model and choosing an appropriate target utility. The fact that the solution of any HS-problem is efficient, implies that the solution of any AM-problem is also efficient.[24]

The above result can be summarized graphically. Let NR, TDR and Q, respectively, be the value of the objective function, the total differential rent, and the shadow income subsidy at the solution of each HS-problem. Then, a simple calculation reveals that the following relationship holds (under any parameter values of u and M):

$$\mathrm{NR} = \mathrm{TDR} - MQ. \tag{1.43}$$

The relationship between the two curves NR and TDR is depicted in Figure 1.8. Since Q is increasing in u ($d\hat{Q}(u)/du > 0$), before

[24] This conclusion is, of course, hardly surprising in light of traditional welfare economic theory. Nevertheless, it is worth reconfirming in the context of the model with continuous space.

they intersect at point A the difference between the two curves is decreasing in u; after point A, the difference is increasing in u. Given an income subsidy (or tax) Q, the equilibrium utility $u^*(Q)$ at the solution of $AM(Q)$-problem can be obtained as in Figure 1.8. In particular, Q equals zero at point A. Hence, this point corresponds to the market equilibrium for the closed city model under absentee land ownership in subsection 1.2 (case 1). Next, at point B which exists uniquely since curve NR(u) is continuous and $\lim_{u \to \infty} \text{NR}(u) = -\infty$, we have $\text{NR} = 0$ and hence $Q = \text{TDR}/M$. Hence, point B corresponds to the equilibrium of the public ownership market model of (1.22) under fixed population. Thus, we can conclude that *there exists a unique equilibrium for the closed city model under public land ownership in subsection* 1.2 (*case* 3).

Thus far, we have been concerned with the effect of change in parameter u on the solution of the HS-problem, while population M is kept constant. Next, let us examine the effect of change in parameter M on the solution of the HS-problem, while the target utility level is kept constant at u. As depicted in Figure 1.9, the net revenue curve, NR(M), is continuous and strictly concave in M, and $\lim_{M \to \infty} \text{NR}(M) = -\infty$ (see Fujita [46]). Let $\hat{u}$ be the supreme utility level determined by the relation, $\Psi(0, \hat{u}, 0) = R_A$. We can see that if $u < \hat{u}$, then $\partial\text{NR}/\partial M > 0$ at $M = 0$, and hence intersection B exists uniquely. Since $\text{NR} = 0$ at point B, from (1.43) $Q = \text{TDR}/M$.

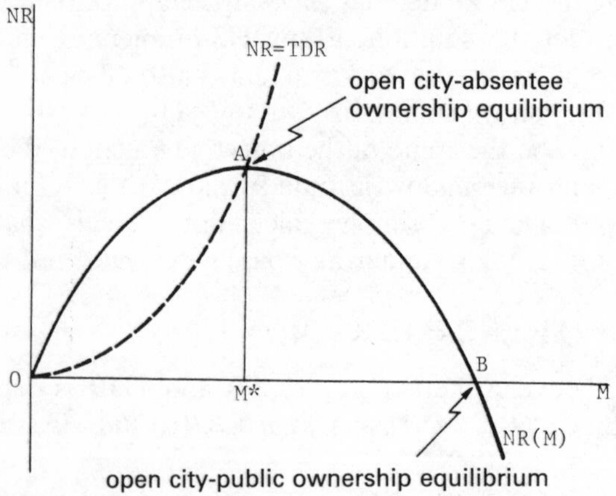

FIGURE 1.9 Relationship between optimal solutions and equilibrium solutions (u fixed).

Utilizing this result, it is not difficult to confirm that if M is given at point B, optimality conditions (1.43)–(1.36) coincides with the equilibrium conditions for the public ownership model of (1.22) under free migration. Hence, we can conclude that *if the national utility level is lower than* $\hat{u}$, *there exists a unique equilibrium with positive population for the open city model under public land ownership* (*case* 4). Finally, point A in Figure 1.9 corresponds to the optimal population for the city developer who aims to maximize the net revenue. At this point, $\partial \mathrm{NR}/\partial M = 0$. Applying the envelope theorem to (1.43), we have $\partial \mathrm{NR}/\partial M = -Q$. Hence, at point A, $\mathrm{NR} = \mathrm{TDR}$ and $Q = 0$. Utilizing this result, it is not difficult to confirm that point A corresponds to the equilibrium of the open city model under absentee land ownership (case 2).

1.4. Some extensions

Having mastered the basic model (1.1), it is appropriate in this subsection to incorporate some of the important factors which we have previously neglected.

A. Time cost of commuting. Although we have not explicitly considered the time cost of commuting, it is so important in practice that some authors (e.g., Beckmann [22]; Henderson [61], Hochman and Ofek [64]) considered only that cost, neglecting the pecuniary cost. In order to examine the effects of pecuniary cost and time cost on residential choice, let us consider the following model of residential choice, which is a simplified version of Yamada [155][25]

$$\max_{r,z,s,t_l,t_w} U(z, s, t_l), \quad \text{subject to} \tag{1.44}$$

$$z + R(r)s + ar = Y_N + wt_w, \quad \text{and} \quad t_l + t_w + br = \bar{t},$$

where z, s, and $R(r)$ are the same as before; t_l represents the leisure time, t_w the time for working, b the commuting time per distance, $\bar{t}$ the total available time, Y_N the nonwage income, w the wage rate, and a the pecuniary commuting cost per distance. That is, the

[25] In Yamada [155], other factors such as disutility of working time and commuting time and environmental external effects are also considered. Note that here the household can freely choose the length of working time. For the case where maximum working length is constrained, see Yamada [155] and Moses [92].

household faces both a budget and a time constraint. From the time constraint, $t_w = \bar{t} - t_l - br$. Thus, assuming that t_w is always positive at the optimal choice, the above residential choice model can be restated as follows:

$$\max_{r,z,s,t_l} U(z, s, t_l), \text{ subject to } z + R(r)s + wt_l = I(r), \tag{1.45}$$

where $I(r) \equiv Y_N + I_w(r) - ar$, and $I_w(r) \equiv w(\bar{t} - br)$. Note that wage rate w now represents also the *unit price of leisure time*. We call $I(r)$ the *potential-net-income* at r. With usual assumptions on the utility function, define the bid rent function by

$$\Psi(r, u) = \max_{s,t_l} \frac{I(r) - Z(s, t_l, u) - wt_l}{s}, \tag{1.46}$$

where $Z(s, t_l, u)$ is the solution of $U(z, s, t_l) = u$ for z. As before, we assume that land s is a normal good. Then, it is not difficult to show that the locational effect of nonwage income is the same as before: other aspects being equal, *households with higher nonwage incomes locate farther from the CBD than households with lower nonwage incomes.*

In order to examine the effect of wage rate w, denote the ordinary demand function for land by $\hat{s}(I, R, P_l)$ where P_l is the price of leisure time. And, define

$$\eta \equiv \frac{\partial \hat{s}}{\partial I}\frac{I}{s}, \qquad \varepsilon \equiv \frac{\partial \hat{s}}{\partial P_l}\frac{P_l}{s}. \tag{1.47}$$

By definition, η is the potential-net-income elasticity of lot size, and ε the cross-elasticity of lot size to the price of leisure time. Then, we can show that[26]

$$-\frac{\partial \Psi'}{\partial w}\bigg|_{d\Psi=0} \gtreqless 0 \quad \text{as} \quad f(r, w) \equiv \frac{1}{1 + (a/bw)} - \left(\frac{I_w(r)}{I(r)}\eta + \varepsilon\right) \gtreqless 0, \tag{1.48}$$

where $\Psi' \equiv \partial\Psi/\partial r$.

[26] For the derivation of (1.48), see Fujita [46, Ch. 2]. In detail, the mathematical operation, $\partial\Psi'/\partial w\,|_{d\Psi=0}$, means $\partial\Psi'/\partial w_{\Psi(r,u|w)=\text{const}}$. That is, by *keeping the value of* $\Psi(r, u \mid w)$ *constant,* we examine how the slope of bid rent curve changes at each (r, u) when parameter w is changed. Recalling the definition of relative steepness of bid rent functions, we can conclude that if $-\partial\Psi'/\partial w\,|_{d\Psi=0}$ is positive (negative) at every point (r, u), then the bid rent function Ψ becomes steeper (less steep) as w increases.

Consider the case of Beckmann [22] and Hochman and Ofek [64], where both Y_N and a are assumed to be zero. In this case, $f(r, w) = 1 - (\eta + \varepsilon)$. Thus, from (1.48), if $\eta + \varepsilon < 1(\eta + \varepsilon > 1)$, the bid rent function becomes steeper (less steep) with increasing wage rate. Thus, recalling Rule 1.3, we can conclude that *in the case of the pure wage earner with negligibly small pecuniary transport cost, if the wage elasticity of lot size, $\eta + \varepsilon$, is less (greater) than unity, the equilibrium location of the household moves in (moves out) with increasing wage rate.* This result was first obtained by Hochman and Ofek [64]. In Japan, for example, pecuniary commuting costs are often paid by employers ($a = 0$). Therefore, if condition $\eta + \varepsilon < 1$ obtains, which is most likely the case, this proposition explains the general tendency in most Japanese large cities for wealthy households to live closer to the CBD.

When pecuniary commuting costs are not neglibible (as in the U.S.A.), we cannot apply the above proposition; and hence we reconsider relation (1.48). Again, consider the case of the pure wage earner. Then, in a realistic range of parameters ($a, b, w, \bar{t}, r$), it is quite safe to use the next approximation of function $f(r, w)$ (i.e., $I_w(r)/I(r) \doteqdot I_w(0)/I(0) = 1$):

$$f(r, w) \doteqdot f(w) \equiv 1/(1 + (a/bw)) - (\eta + \varepsilon). \qquad (1.49)$$

Recent empirical studies in the U.S.A. indicate that wage elasticity of lot size, $\eta + \varepsilon$, may be considerably less than unity.[27] In this case, from (1.48), the slope of bid rent function, $-\Psi'$, changes as depicted in Figure 1.10, where $\hat{w} = (a/b)((\eta + \varepsilon)/(1 - (\eta + \varepsilon))$.[28] Thus, from Rule 1.3, we can conclude that *increases in wage rate first move the equilibrium location away from the city center*; *but, beyond wage rate $\hat{w}$, additional increases retract the household location.* The result is that both very low wage earners and very high wage earners tend to reside near the city center; middle wage earners gravitate towards the suburbs. This is consistent with what has been observed of large cities in the U.S.A. For example, the

---

[27] Estimated values for the realized-gross-income elasticity of housing are 0.75 by Muth [94], 0.5 by Carliner [30], and 0.75 cited by Polinsky [110]. Wheaton [150] estimates the realized-gross-income elasticity of land to be 0.25. Since the value of ε will be close to zero, these empirical studies suggest that $\eta + \varepsilon$ may be considerably less than one.

[28] If we use the same parameter values from Altman and Desalvo [3], which studies the U.S. cities for the period 1960 to 1975, we have $\hat{w} = 11{,}270$/year. worker.

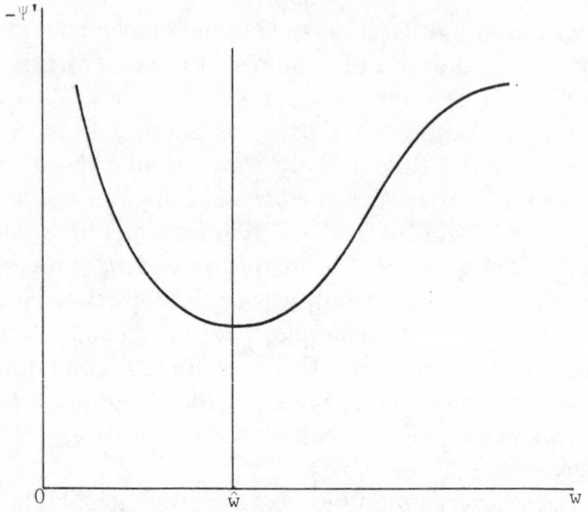

FIGURE 1.10 Effect of wage rate on slope of bid rent function.

behavior of the curve in Figure 1.10 is consistent with the estimate of slopes of bid rent curves in San Francisco by Wheaton [150].[29]

B. Family structure. The model (1.44) can also be extended to encompass the effects of family structure on the locational decision. Following Beckmann [21], we assume that the family structure of a household is characterized by two parameters: d, the number of dependent members, and n, the number of working members in the household. Thus, model (1.44) becomes:

$$\max_{r,z,s,t_l,t_w} U(z, s, t_l; d, n), \qquad \text{subject to} \tag{1.50}$$

$$z + R(r)s + nar = Y_N + nwt_w, \text{ and } t_l + t_w + br = \bar{t}.$$

Note that the first constraint represents the *family* budget, and the second the time constraint for *each* working member. This can be rewritten as follows:

$$\max_{r,z,s,t_l} U(z, s, t_l; d, n), \text{ subject to} \tag{1.51}$$

$$z + R(r)s + nwt_l = I(r, n),$$

[29] Here in order to explain the general pattern of household location observed in the U.S.A., we used model (1.44). LeRoy and Sonstelie [77] present an alternative model which introduces multiple transport modes.

where $I(r, n) \equiv Y_N + nw(\bar{t} - br) - nar$. Thus, the bid rent function is now given by

$$\Psi(r, u) = \max_{s, t_l} \frac{I(r, n) - Z(s, t_l, u; d, n) - nwt_l}{s}, \qquad (1.52)$$

where $Z(s, t_l, u; d, n)$ is the solution of $U(z, s, t_l; d, n) = u$ for z.

As an example, let us consider the case of the next log-linear utility function:

$$U(z, s, t_l; d, n) = h\alpha \log(z/h^k) + h\beta \log(s/h^\delta) + n\gamma \log t_l + d\lambda \log \bar{t}, \qquad (1.53)$$

where $h = d + n$, and each of α, β, γ, λ, k and δ is a positive constant. Combined (1.50) and (1.53), the model represents an extension of Beckmann [21].[30] A simple calculation yields that

$$-\frac{\partial \Psi}{\partial r} = \frac{\alpha + \beta + (n/h)\gamma}{\beta} \frac{a + bw}{(Y_N/n) + w(\bar{t} - br) - ar} \Psi(r, u). \qquad (1.54)$$

Since $h = d + n$, we can see that the bid rent function becomes less steep with increasing number of dependents: $-\partial\Psi'/\partial d\,|_{d\Psi=0} < 0$. Thus, from Rule 1.3, we can conclude that *the more dependents a household has, the farther is its equilibrium location from the CBD.* Next, in the case of pure wage earners ($Y_N = 0$), the slope of the bid rent function depends only on the ratio, n/h. Again, from Rule 1.3, we can conclude as follows. *In the case of pure wage earners, locations may be ranked by the households' commuter-family size ratio, n/h: the smaller the ratio, the farther the location from the CBD.* Similarly, we can conclude from (1.54) that *in the case of pure wage earners with no dependents ($d = 0$, and hence $n/h = 1$), locations are independent of family size (i.e., the number of commuters).* These conclusions, first obtained by Beckmann [21], fit well with many casual observations from cities in the U.S.A.[31]

C. Housing service consumption. In the basic model of (1.1), it is implicitly assumed that each household manages the construction of

[30] In Beckmann [21], the pecuniary transport cost is assumed to be zero ($a = 0$), and t_w is fixed.

[31] Although these conclusions result from a log-linear utility function, Fujita [46, Ch. 2] presents more general results in terms of the original model, (1.50).

its house by itself. There is, however, another class of models, originated by Muth [93], in which households are assumed to consume an aggregate commodity called the *housing service.* That is, each household behaves as:

$$\max_{r,z',q} U(z', q), \quad \text{subject to} \quad z' + P(r)q = Y - T(r), \qquad (1.55)$$

where $P(r)$ is the unit price of housing service q at location r, and z' represents the amount of composite consumer good excluding housing service. In turn, the housing industry produces the housing service with production function $F(s, k)$ from land s and non-land input k. That is, the housing industry behaves as:

$$\max_{s,k} P(r)F(s, k) - R(r)s - P_k k, \quad \text{at each } r, \qquad (1.56)$$

where $R(r)$ is the land rent at r, and P_k is the unit price of non-land input which is assumed to be exogenously given independently of location. If we assume, as in Muth [93], that the housing production function is of constant returns to scale, then in equilibrium, $P(r)F(s, k) = R(r)s + P_k k$. Thus, Muth model is equivalent to the following reduced form model in which each household chooses land and non-land input by itself:

$$\max_{r,z',s,k} U(z', F(s, k)), \text{ subject to } z' + P_k k + R(r)s = Y - T(r). \qquad (1.57)$$

With appropriate changes in notation and utility function, this is mathematically a special case of the basic model, (1.1).[32] However, since Muth's model is richer and more specific in context than the basic model, it can provide more specific results. For example, under the assumption of price elasticity of compensated demand for housing service equal to minus one, Muth derived a negative exponential housing price curve. Further, with a Cobb-Douglas housing production function, he showed that the population density curve is also negative exponential.[33]

[32] That is, define $U(z, s) = \max_{z',k} \{U(z', F(s, k)) \mid z' + P_k k = z\}$.

[33] For more detailed discussion on housing, see the survey article by Muth and Quigley in this series. See, for example, Niedercorn [95] for further discussion on negative exponential models of urban land use.

D. Transport network. Although we have assumed that the transport cost is the same in every direction from the CBD, more realistic, non-uniform transport networks have been considered in Muth [93, Ch. 4] and Anas and Moses [9]. More realistic land use patterns arise with the introduction of non-uniform transport networks. However, as long as transport congestion is absent, theoretically no essential change is necessary in our basic model in order to handle the case of the non-uniform transport network. That is, let us arbitrarily fix a radial straight line from the CBD, and let r measure the distance from the CBD on that line. Let $L(r)$ be the amount of the land on the *iso-transport contour* passing each distance r on that line. Then, without any change, all the previous results also apply for this case.

E. Multicentric cities. Although we have assumed a single concentration of employment and shopping, the CBD, many works have introduced multiple, *prespecified* centers of employment or shopping. See, for example, Muth [93], Papageorgiou and Casetti [105], Hartwick and Hartwick [55], Odland [96], White [153], and Romanos [117].

2. RESIDENTIAL LAND USE WITH EXTERNALITIES

In this section, we introduce externalities into the basic theory of residential land use. Our focus is on the spatial aspects of externalities, since other aspects receive in-depth treatment elsewhere.[34] The basic model (1.1) is now generalized as follows:

$$\max_{r,z,s} U(z, s, E(r)), \text{ subject to } z + R(r)s = Y - G(r) - T(r), \tag{2.1}$$

where $E(r)$ represent the level of externalities at distance r from the CBD,[35] and $G(r)$ is the tax per household at r. We may call (2.1) the *externality model* of residential choice. The bid rent function

[34] See survey articles by Kanemoto and Schweizer in this series.
[35] In general, $E(r)$ could be a vector. In most of the following discussion, however, we consider one kind of externality at a time. Hence, unless otherwise noted, $E(r)$ is a single number.

becomes

$$\Psi(r, u) \equiv \Psi(r, u; E(r)) = \max_{s} \frac{Y - T(r) - G(r) - Z(r, u, E(r))}{s}, \qquad (2.2)$$

where $Z(r, u, E(r))$ is the solution of $U(z, s, E(r)) = u$ for z.

$E(r)$ represents externalities from public facilities in subsection 2.1, and neighborhood externalities in subsection 2.2. In subsection 2.3, transport cost is subject to traffic congestion with $E(r)$ being omitted.

2.1. Local public goods and location of public facilities

The famous paper by Tiebout [145] marks the beginning of abundant studies of the efficient provision of local public goods (among communities). To a great extent these studies have ignored the spatial dimension in modeling local public goods. Here, we shall focus on studies which do consider the spatial dimension and land market. Such studies fall into 3 groups which are differentiated in accordance with the spatial distribution of public goods.

The first group concerns pure public goods, which assumes the externality level, $E(r)$, to be the same everywhere in the city. Barr [18] and Yang [156] examined the impacts of pure public goods on residential density, land rent and city size. The issue of how the efficient level of pure public goods may be determined in a competitive spatial economy was demonstrated by Kanemoto [70] by applying Margolis' "fiscal profitability principle." He showed that in a system of a fixed number of small, open cities the optimal provision of a public good (local to each city) will be achieved if each city government behaves so as to maximize the total differential rent (TDR) minus the cost of public good (K). And, if the number of cities is optimal, it holds that

$$\text{TDR} = K, \qquad (2.3)$$

which represents the so-called *Henry George Theorem* or the *Golden Rule of local public finance.* The same result was obtained by Arnott and Stiglitz [16] and Arnott [12] in the context of optimal allocation.[36] Helpman and Pines [59] presents a similar analysis on the optimal public investment for amenity improvements in a system of cities. An interesting work by Brueckner [25] characterized the spatial equilibrium with a pure public good whose output is chosen by majority voting.

The second group of studies concerns spatially continuous local public goods, which assumes $E(r)$ to depend on the amount $X(r)$ of public survives *provided* at distance r and the number $n(r)$ of households at r:

$$E(r) = f(X(r), n(r)). \tag{2.4}$$

If $\partial f/\partial n = 0$, X represents noncongestible public services; if $\partial f/\partial n < 0$, congestible. Schuler [125, 126] and Helpman, Pines and Borukhov [60] showed that when the optimal allocation of households is achieved through the competitive market (while public services are provided by the city government), the optimal tax is the sum of a lump-sum tax g and a location tax $l(r)$:

$$G(r) = g + l(r), \tag{2.5}$$

where $l(r)$ is zero in the case of noncongestible public services. Kanemoto [70] showed that a system of competitive land developers will achieve the optimal allocation of public services and households if at each distance a developer taxes the households and maximizes the net land rent plus tax minus the costs of providing public services. These studies assume that the provision of public services does not require land. Yang and Fujita [157] treated open space as a spatially continuous public good, and studied the competition in the land market between the public sector and the household sector. Although it was briefly examined by Fisch [36] and Yang and Fujita [157], the study of spatially continuous public goods with spillover effects is largely left for the future.

The third group of studies concerns public goods which are provided at discrete location because of scale economies. In practice, many important public goods (e.g., schools, hospitals) belong to this class. Although the problem of locating public

---

[36] For further discussion of local public goods and Henry George Theorem, see survey articles by Wildasin and Schweizer in this series.

facilities has received considerable attention, most of the existing works concern applied public facility location model.[37] With the exception of Sakashita [120], few theoretical land use models with (point-pattern) public facilities have been developed. Sakashita presents an insightful analysis of the optimal location of public facilities under the influence of the land market. Although Fujita [44] achieved a two-dimensional extension of a part of Sakashita's work, further theoretical developments are largely left for the future.

2.2. Neighborhood externalities

In model (2.1), let us now assume that $E(r)$ represents neighborhood externalities. These have been described in two classes of models, *crowding models* and *racial models*.

In crowding models, $E(r)$ represents local environmental amenities and is set equal to the average neighborhood lot size (the reciprocal of household density) at distance r. Richardson [113] suggested that effects of crowding may result in a positive gradient at the equilibrium. That is, if the household's preference for low density is sufficiently strong, increasing average lot size $E(r)$ may cause a rising land rent curve $R(r)$ nearer the CBD. This interesting conjecture was, however, proved to be unlikely. Actually, Grieson and Murray [51] showed that a positive rent gradient equilibrium is either unstable or arises only if land is an inferior good, and that it will never occur under efficient allocations. Tauchen [141] showed that the land rent gradient is always negative if the utility function is strictly quasi-concave and if individual and average neighborhood lot size are substitutes. Further, Scotchmer [132] showed that in the presence of crowding effects, if (decentralized) land developers maximize the land rent revenue (per unit land) at each distance with respect to lot size, then the competitive equilibrium will be efficient.

Three types of models have been proposed to capture the effects of racial externalities. These models have mostly been developed in the U.S.A. where racial problems have been a prime concern. The models assume that whites have an aversion to living near blacks

[37] For recent works on public facility location, see Thisse and Zoller [146] and the review article by Lea [76].

while blacks may or may not prefer to live close to whites. By further assuming that all blacks are identical and all whites are identical, model (2.1) may be rewritten as follows:

$$\max_{r,z,s} U_i(z, s, E_i(r)), \text{ subject to } z + R(r)s = Y_i - G_i(r) - T_i(r), \tag{2.6}$$

where $i = B, W$. And, the bid rent function (2.2) becomes:

$$\Psi_i(r, u) \equiv \Psi_i(r, u; E_i(r)) = \max_s \frac{Y_i - G_i(r) - T_i(r) - Z_i(r, u, E_i(r))}{s}. \tag{2.7}$$

The so-called *border models* developed by Bailey [17] and Rose-Ackerman [118, 119] assume *a priori* that blacks and whites are completely segregated with blacks in the city center. Here, $E_i(r)$ is a non-increasing function of the distance from the border, b:

$$E_i(r) = f_i(|r - b|),\ df_i/d\,|r - b| \leqq 0,\ i = B, W. \tag{2.8}$$

If $Y_B < Y_W$, $T_B(r) = T_W(r)$, and $G_B(r) = G_W(r) = 0$, then, as shown by Rose-Ackerman (1975), the effects of racial prejudice can be demonstrated by Figure 2.1. If the city is free of prejudice

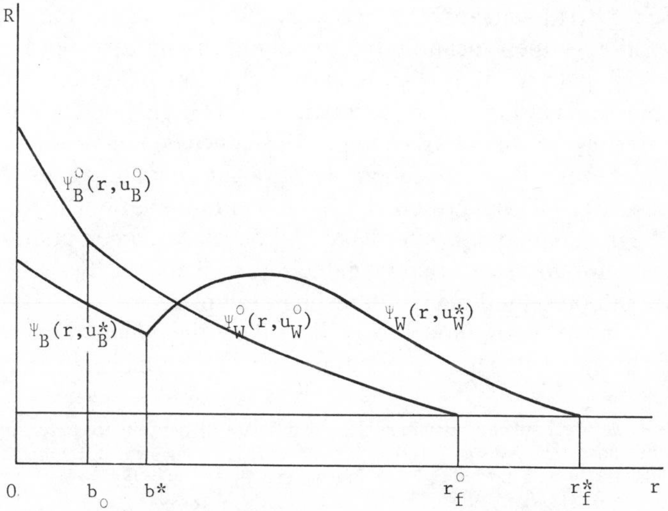

FIGURE 2.1 Effects of whites' prejudice against blacks.

($\partial U_B/\partial E = \partial U_W/\partial E = 0$), the city has a land rent curve such as that given by $\Psi_B^0(r, u_B^0) - \Psi_W^0(r, u_W^0)$. As we have seen in Section 1.1, blacks with a lower income have a steeper equilibrium bid rent curve.[38] When blacks are indifferent to location of whites ($\partial U_B/\partial E_B = 0$) but whites are repelled from blacks ($\partial U_W/\partial E_W < 0$), the city has land rent curve shown by $\Psi_B(r, u_B^*) - \Psi_W(r, u_W^*)$ in Figure 2.1. With the introduction of whites' aversion to blacks, both the city boundary and the border between blacks and whites move outward, the land rent curve becomes flatter (possibly, a positive gradient near b), and blacks' utility rises ($u_B^0 < u_B^*$). This pattern of land use in the prejudiced city is not Pareto-efficient. One way to achieve an efficient land use in the prejudiced city is to impose an appropriate location tax, $G(r)$, on blacks. This, however, may be morally and politically infeasible.

Border models were criticized by Courant and Yinger [31] for two main reasons. First, they *a priori* assume a completely segregated pattern with blacks in the city center. Second, if some blacks have higher incomes than some poor whites, then this completely segregated pattern can not be sustained within the context of border models. The second type of model referred to as the *local externality model* was proposed by Yinger [159] and Schnare [124].[39] Here, $E_i(r)$ equals the proportion of black households at each distance r. This means that households care about the racial composition of their location, but not about that in other locations. The racial pattern of land use in the city can be endogenously determined. By using Yinger's model, Kern [73] obtained a number of interesting results. For example, he concluded that whenever preference for white neighbors is stronger among whites than among blacks, the integrated equilibrium is unstable, but there is a stable segregated equilibrium. King [74] developed a computational algorithm for local externality models (based on the Scarf Algorithm), and presented many interesting examples.

The third type of model may be called the *global externality model*. In these models, the total amount of externalities received,

[38] Here, we are implicitly assuming that in the case of unprejudiced city, blacks and whites have the same utility function, $U(z, s)$. And, in order to compare the unprejudiced city with prejudiced one, we assume that $E_i(\infty) = 0$ for $i = B$ and W, and $U_B(z, s, 0) = U_W(z, s, 0) = U(z, s)$.

[39] Yinger [159] called it the local amenity model.

for example, by a white is assumed to be a weighted sum of externalities from all blacks in the city, where weights are a decreasing function of distance. Examples are Yellin [158], Papageorgiou [103, 104], Kanemoto [70, Ch. 6] and Ando [10, Ch. 5]. Compared to the other racial externality models, the global model is both most general and analytically most complex. The systematic analysis of solution characteristics (both for equilibrium and optimal) for global externality models such as existence, uniqueness, symmetricity and stability is largely left for the future.[40]

2.3. Transport congestion and land use for transport

So far, we have assumed that the city is free of transport congestion and hence transport cost $T(r)$ can be given as an exogenous function of distance r. In fact, transport congestion is probably the most important kind of externality in cities, and has received considerable study.[41] Optimal urban land use with transport congestion was first studied by Strotz [139] within the context of monocentric discrete rings. Solow and Vickrey [137] examined the optimal allocation of land between business activity and transportation in a long-narrow city. Mills and de Ferranti [84] was the first to study the optimal allocation of land between housing and transport in a standard, continuous monocentric city. In the following, we will extend the basic theory of Section 1 to include transport congestion. For simplicity, we assume that automobiles are the only mode of transportation, and that the only cost of building roads is the opportunity cost of land. After studying the optimal allocation of land between housing and transport, we can then show that a location tax is necessary for an efficient allocation through the competitive market.

Let $N(r)$ denote the number of households living beyond distance r, and $L_T(r)$, the amount of land devoted to transport use at distance r. Assuming that one member of each household commutes to the CBD, $N(r)$ equals the number of commuters passing through distance r. Following Kanemoto [69], the marginal transport cost at

[40] For further discussion on racial models, see the survey article by Kanemoto in this series.

[41] For historical account of studies on urban land use with transport congestion, see, for example, Goldstein and Moses [49] and Richardson [112, Ch. 4].

each distance r is assumed to be a function of the traffic-land ratio, $N(r)/L_T(r)$, and it is denoted by $g(N(r)/L_T(r))$. Then, the transport cost $T(r)$ is given by

$$T(r) = \int_{r_c}^{r} g(N(r)/L_T(r))\, dr, \tag{2.9}$$

where r_c is the fringe distance of the CBD. We assume r_c is positive, but neglect the transport cost within the CBD. The marginal transport cost function, $g(x)$, is assumed to be increasing and strictly convex in x, and $g(0) > 0$, and $\lim_{x\to\infty} g(x) = \infty$. If we arbitrarily specify a target utility level, u, then the net revenue NR can be given as before by Eq. (1.32). Thus, we obtain the Herbert-Stevens problem, $\mathrm{HS}_T(u)$, with transport congestion:

$$\max_{L_T(r), n(r), s(r), r_f} \mathrm{NR} = \int_{r_c}^{r} \{(Y - T(r) - Z(s(r), u) - R_A s(r))n(r) - R_A L_T(r)\}\, dr, \tag{2.10}$$

subject to (2.9) and the following constraints,

$$L_T(r) + n(r)s(r) \leqq L(r), \tag{2.11}$$

$$\int_{r}^{r_f} n(r)\, dr = N(r), \quad \text{and} \quad N(r_c) = M, \tag{2.12}$$

where M is as usual the total number of households in the city.

Let Q denote an income subsidy per household and $l(r)$ denote the location tax per household at distance r. As in (1.33), let us define the household bid rent function as follows:

$$\Psi(r, u; Q - l(r)) = \max_s \frac{T + Q - l(r) - T(r) - Z(s, u)}{s}. \tag{2.13}$$

The corresponding bid-max lot size is $S(r, u; Q - l(r))$. Define the *bid rent function* $\Psi_T(r)$ *of the transport sector* as the marginal benefit of land for transport:

$$\begin{aligned}\Psi_T(r) &= [-\partial g(N(r)/L_T(r))/\partial L_T]N(r)\\ &= g'(N(r)/L_T(r))(N(r)/L_T(r))^2.\end{aligned} \tag{2.14}$$

Then, the necessary and sufficient conditions for optimal allocation are given by (2.9), (2.11), (2.12), and by the following conditions:[42]

$$R(r) = \max\{\Psi(r, u; Q - l(r)), \Psi_T(r), R_A\}, \tag{2.15}$$

$$R(r) = \Psi(r, u; Q - l(r)) \quad \text{if} \quad n(r) > 0, R(r) = \Psi_T(r) \quad \text{if} \quad L_T(r) > 0, \tag{2.16}$$

$$s(r) = S(r, u; Q - l(r)) \quad \text{and} \quad n(r) = (L(r) - L_T(r))/S(r, u; Q - l(r)) \quad \text{for} \quad r \leqq r_f, \tag{2.17}$$

$$l(r) = \int_{r_c}^{r} g'(N(r)/L_T(r))N(r)/L_T(r)\,dr \quad \text{for} \quad r \leqq r_f. \tag{2.18}$$

Meanings of conditions (2.15), (2.16) and (2.17) are obvious. In order to see the meaning of (2.18), note that when an additional commuter living outside r commutes through r, this person causes the congestion cost of $g'(N(r)/L_T(r))/L_T(r)$ for every existing commuter passing through r. Thus, (2.18) says that an *optimal location tax* $l(r)$ *equals the total congestion cost caused by a person commuting from distance* r *to the CBD.* From (2.18), *location tax* $l(r)$ *is everywhere positive and increasing in* r. Figure 2.2 depicts a possible optimal land use configuration.[43]

Next, in the context of the competitive market, let us assume that the residential choice behavior of each household is given by

$$\max_{r,z,s} U(z, s), \text{ subject to } z + R(r)s = Y + Q - l(r) - T(r),$$

where Q and $l(r)$ an income subsidy and location tax specified by the city government. And, assume that the transport sector choose the amount $L_T(r)$ of land for transport at each distance so as to set the marginal benefit of land for transport equal to the market land rent:

$$[-\partial g(N(r)/L_T(r))/\partial L_T]N(r) = R(r).$$

Then, we can see from (2.18) that the market allocation of land is efficient only if the city government imposes location tax $l(r)$ which satisfies relation (2.18). Robson [116] and Kanemoto [70, Ch. 4]

---

[42] For derivation of these optimality conditions, see Ando [10] or Fujita [46].

[43] Figure 2.2 is based on the following specifications: $L(r) = \theta r$, $U(z, s) = \alpha \log z + \beta \log s$, $g(N/L_T) = a(N/L_T)^b$. It is also possible that $L_T(r) = L(r)$ (i.e. saturation) near the CBD fringe.

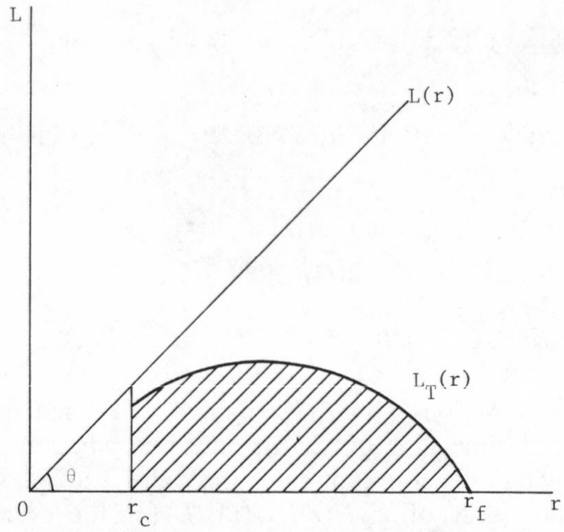

FIGURE 2.2 Optimal land use configuration.

showed, by using a Cobb-Douglas utility function and a polynomial marginal transport cost function, that the market city (with no location tax) is more spread out than the optimal city.[44]

3. GENERAL EQUILIBRIUM MODELS

This section concerns general equilibrium models of urban land use, in which the location of households and firms is determined simultaneously. Indeed, overall, the task of such models is to generate structured space from a void space. With that in mind, the assumption of a monocentric city employed in the previous sections is not appropriate here. It has both theoretical and empirical drawbacks. Theoretically, both the spatial organization of land uses and whether a city is monocentric or not should be explained within the framework of the model itself, without assuming any *a priori* locations. Such a model might well yield a monocentric configuration given a particular parameter set. But it could also yield other

[44] The market city means the equilibrium land use in the case of public land ownership, and the optimal city is the case of $NR = 0$. For further discussion on land use for transport see the survey article by Kanemoto in this series.

configurations, and herein lies the potential explanatory power. This argument is strengthened by results of empirical work (e.g., Kemper and Schmenner [72], Mills [81], Odland [97] suggesting that the monocentric configuration is but one of many spatial patterns a city can assume. In this section, we hope to show that many types of general equilibrium models of urban land use can be developed and that these models open new perspectives to urban economics.[45]

3.1. Prototypes of general equilibrium models

A general equilibrium model of urban land use has a *trivial solution* when it generates no spatial agglomeration and the land use pattern follows a uniform distribution of all activities. Not only are trivial configurations rarely observed, but theoretically, they are not interesting. To construct a meaningful and interesting model, then, with nontrivial solutions, we shall begin by recalling the following result from Starrett [138]. Suppose we have an economy in which the following assumptions are met:

a1. (*no relocation cost*)*:* All agents are free to choose locations.

a2. (*homogeneous space*)*:* All immobile resources are evenly distributed over the space, and hence utility functions of households and technology of firms are independent of location.[46]

a3. (*closed economy*)*:* There is no trade with the rest of the world.

a4. (*perfect markets*)*:* There exist complete markets for all goods at all locations.

Then, from Starrett (138), we can conclude as follows:

Spatial Impossibility Theorem: Under assumptions a1–a4, there is no competitive equilibrium with a positive total transport cost, so the only possibly competitive equilibrium is the one in which no good or person is transported (with positive cost).

By *competitive equilibrium,* we mean the market equilibrium in which all agents take prices as given. Under assumptions a1–a4, an urban configuration can be in equilibrium and no transport is involved only when all activities are uniformly distributed. Therefore, it follows that if assumptions a1–a4 are upheld, there exists

[45] For more detailed discussion, refer also to Fujita [43].

[46] This assumption does not imply a uniform transport plain. Some non-uniform transport networks may exist as explained in the next footnote.

either a trivial solution, or no (price-taking) competitive equilibrium. In short, the spatial impossibility theorem says that the smooth market mechanism alone cannot generate spatial agglomeration of activities. Given that the free mobility of agents is an essential feature of the long-run problem, Mills [82, 83] and Kanemoto [70] have suggested the following possible causes of spatial agglomerations in a city:

b1. uneven distribution of natural resources.

b2. proximity to economical transportation with the rest of the world.[47]

b3. increasing returns to scale or indivisibility.

b4. externalities or public goods.

With respect to the first factor, it is worthwhile to note that the distribution of natural resources is not generally important within the context of urban land use, so the assumption of homogeneous space can still be maintained. With respect to the third factor, one should note that at the individual agent level, increasing returns to scale are incompatible with price-taking competitive equilibrium. With these arguments in mind, meaningful general equilibrium models fall into three categories:

A. Port City Model: perfectly competitive equilibrium model with interurban trade.

B. Spatial Externality Model: competitive equilibrium model with spatial externalities.

C. Imperfect Competition Model with scale economies.

In the following pages, we will discuss each of these prototypes. As will be shown, each prototype modifies the assumptions underlying the Spatial Impossibility Theorem in some way. Since Type A has well studied, emphasis will be on Types B and C.[48]

[47] In the context of intraurban spatial structure, proximity to intraurban transportation is also important. As noted by Mills [82, pp. 18–19], however, intraurban transport access is a determinant of urban spatial structure if there are some reasons for intraurban trade, but is not itself a reason for intraurban trade.

[48] In closing this subsection, it is appropriate to mention the famous *spatial assignment problem* by Koopans and Beckmann [75]. The problem is to assign n agents to n plots so as to maximize the net revenue. It is assumed that each agent is indivisible, that no plot can be occupied by more than one agent, and that each agent requires transactions of some positive amount of intermediate goods with other agents. Assumptions a1 and a3 of the Spatial Impossibility Theorem are also satisfied. Then, they conjectured that no integer assignment will be sustainable by a competitive price system. Following an unsuccessful attempt by Hartwick [53],

3.2. Port city model: Type A

Here we will relax assumption a3 concerning the closed economy, and allow the city to trade with the rest of the world through a trade node (port, or station) whose location is fixed *a priori.*[49] No externalities, public goods or scale economies in production exist. In other words, the Type A model describes the perfectly competitive equilibrium structure of a city which is caused by proximity to economical interurban transportation. As expected from traditional welfare economics, the equilibrium allocation in such a city is always efficient. Examples are Mills [80] and [81, Ch. 5], Solow [136], Goldstein and Moses [50], White [153], Schweizer and Varaiya [129, 130], Henderson [61, Ch. 1], Schweizer [128], Karmann [71], and Sullivan [140].[50]

Most port city models are of the two-sector variety (households and firms producing exporting goods). Of these, most focus only on monocentric spatial structure, in which the production area is surrounded by residential land.[51] Other fixed-coefficient models by Mills [81, Ch. 5] and Goldstein and Moses [50] suggest that a completely integrated pattern is also possible. And, in the context of more general utility and production functions, still more equilibrium patterns are conceivable.[52] In short, the study of complete

---

Heffley [57] proved the following results: (i) in the case of homogeneous space, no integer assignment is sustainable by any competitive price system, but (ii) in the case of nonhomogeneous space, it is possible for optimal assignments to be sustained by competitive price systems. In fact, since the first three assumptions (a1 to a3) are satisfied in the case of homogeneous space, and since any integer assignment requires a positive transport cost, conclusion (i) immediately follows from the Spatial Impossibility Theorem. Miron and Skarke [86] and Heffley [58] present the further characterization of the spatial assignment problem in terms of game theory. For further discussion of this subject, refer to the survey article by Schweizer in this series.

[49] Equivalently, some models (e.g., Schweizer and Varaiya [129, 130] assume that all final demands are to be traded at a prespecified location. Others (e.g., Solow [136]; Henderson [61, Ch. 2]) require all goods to be traded at the center. White [153] considers multiple exporting nodes.

[50] Since Mills [80] and Sullivan [140] also consider Marshallian external economies in the form of a pure public good, they could be considered as a combination of Type A and Type B models.

[51] Most studies assume zero commuting cost within the CBD, and thus preclude the possibility of other solution patterns.

[52] Although White [153] suggested the possibility of a completely integrated pattern, it was not fully investigated. Capozza [29] presents a good attempt; unfortunately, his analysis is not very precise, and the suggested solution pattern is incorrect.

solution patterns has not yet been achieved. It is important because it may serve to explain job-suburbanization in the context of port city models.[53]

Schweizer and Varaiya [129, 130] and Schweizer [128] have achieved the complete characterization of solution patterns for a general port city model with input-output technology. However, treating households as a pure labor-production sector is unconventional. The task of combining their work with traditional household location models is left for the future.

Although the port city model provides a good starting point, it is not sufficiently general to characterize modern cities. This is because it essentially produces a monocentric land use pattern with no local agglomeration or subcenters. Besides, few modern cities export and import goods exclusively from their centers. Moreover, if there were no exports or imports, the model would generate only the trivial solution.

Finally, it is worth noting that Schweizer, Varaiya and Hartwick [131] extended the Arrow–Debreu model of general equilibrium explicitly specifying "one household at one location." Although we can consider this model as an ultimate generalization of the port city model incorporating an uneven distribution of immobile resources, it suffers essentially the same limitations as the port city model.

3.3. Spatial externality models: Type B

In Type B models, assumptions a1–a3 are satisfied, but a4 is abandoned, and some market imperfections are allowed.[54] The market imperfections relevant to urban land use include *non-price interactions* among agents (e.g., information exchange among firms through face-to-face communication, social interactions among households) and *technological externalities* (e.g., various kinds of technological and informational spillover effects and congestion

[53] An extensive historical study of land use in Chicago by Fales and Moses [35] suggests that land use tended to be quite mixed (between households and firms, and also among different firms) in nineteenth century Chicago. The study of complete solution patterns of a general port city model (with multiple industries) may provide an explanation for that phenomenon.

[54] If we assume that transport cost for trade with the rest of the world is independent of location in the city, we can introduce trade with the rest of the world.

effects). In non-price interactions, agents use non-price signals generated from the spatial allocation of activities (e.g. the average distance to a class of agents), or the spatial allocation itself may act as a signal in determining agents' optimal contact patterns. Technological externalities are also generated from the spatial allocation of activities. We define *spatial externality models* to be those in which such non-price signals or technological externalities influence locational choice. For convenience, we define both technological externalities and the locational effects of non-price interactions as *spatial externalities*. In fact, spatial externalities have been considered a major determinant of spatial structures of modern cities.

A seminal work by Beckmann [23] contains the first development of a spatial externality model. It focuses on how social interactions among households shape equilibrium residential patterns. The utility of each household is assumed to depend on the amount of space occupied and on the *average distance* to all households in the city. Work and shopping trips are ignored. In equilibrium, this residential city exhibits a dispersed population distribution with a single peak. Although Beckmann demonstrated this result in the case of the one-dimensional city, ten Raa [144] generalized it to the case of the n-dimensional city through an elegant application of the Schwartz calculus of distribution. A similar model was also developed by Borukhov and Hochman [24]. Solow and Vickrey [137], Hartwick [54] and O'Hara [100] developed a similar model of business firm location (i.e., office location). In this model, each firm is assumed to have the same number of (or the same probability of) interactions with all other firms in the business district. Thus, the average distance to all other firms is the non-price signal for locational choice. Tauchen and Witte [142, 143] further generalized the Hartwick–O'Hara model so that the contact patterns were endogenously determined; and under a certain specification of contact benefit and facility cost functions, they have demonstrated the existence and uniqueness of the equilibrium by way of a sequential solution of an integral equation. Although these models represent pioneering works, they contain only one sector.

More general Type B models include both households and firms. Examples of two-sector versions are Cappoza [29], Odland [96], Ogawa [98], Ogawa and Fujita [99], Fujita and Ogawa [47], Imai [67], and Fujita [43]. Such models are still in their infancy, limited by a number of *ad hoc* assumptions such as fixed-coefficient technology, exogenous contact patterns, and exogenous firm size. Nonetheless they have been successful in shedding light on some important features of modern cities such as formation of the central *business* district and subcenters, and transition of a monocentric structure to a multicentric structure. In the following, we first present a brief summary of a two-sector model developed by Ogawa, Fujita, and Imai.[55] Then, we discuss future research directions.

Consider a one-dimensional city (i.e., linear city) with two sectors. One sector, made up of business firms, requires non-price interactions, and hence the productivity of each firm climbs with increasing accessibility to others. Assuming that each business firm requires S_b units of land and L_b units of labor, the profit of a business firm at each location may be expressed as

$$\pi(x) = P_0\bar{Q}A(x) - R(x)S_b - W(x)L_b, \qquad (3.1)$$

where $A(x)$ represents the degree of spatial accessibility of location x, of which functional form will be specified later; and $R(x)$ and $W(x)$ are the land rent and wage rate at x.[56] $\bar{Q}$ is the output level of a business firm at the standard condition, $A(x) = 1$, and P_0 is the fixed, unit price of the output. Let the second sector consist of M identical households (or workers) in the city. Each household requires a fixed lot size, S_h. Thus, each household chooses a residential location x and a working site x_w so as to maximize the consumption of composite good z:

$$\max_{x,x_w} z = W(x_w) - t\,|x - x_w| - R(x)S_h, \qquad (3.2)$$

where $W(x_w)$ is the wage rate at x_w, and t is a commuting cost per distance.

[55] Essentially the same model was independently developed by Ogawa and Fujita, and by Imai, which is not surprising since the model represents the analytically tractable simplest two-sector model of Type B.

[56] Alternatively, we may assume that $\pi(x) = P_0\bar{Q} + A(x) - R(x)S_b - w(x)L_b$, where $A(x)$ represents the savings in interaction costs relative to the standard condition, $A(x) = 0$. We can easily see, however, that the two models are mathematically the same after appropriate notational changes. Note also that although we assume all business firms are identical with respect to locational behavior, they may differ in other aspects such as in the services or information they produce.

We define the bid rent function of each household as

$$\Psi(x, u) \equiv \Psi(x, u; W(\cdot))$$
$$= \max_{x_w} (W(x_w) - Z(u) - t\,|x - x_w|)/S_h, \qquad (3.3)$$

where $Z(u)$ is the solution of $u = U(z, S_h)$ for z. Assuming the free entry and exit of business firms, the profit of each firm is zero in equilibrium. Thus, we define the bid rent function of each firm by

$$\Phi(x) \equiv \Phi(x; A(x), W(x))$$
$$= (P_0 \bar{Q} A(x) - W(x) L_b)/S_b. \qquad (3.4)$$

Unknowns to be determined are household distribution $h(x)$, business firm distribution $b(x)$, the land rent curve $R(x)$, wage curve $W(x)$, commuting pattern, and utility level u^*. In equilibrium, we have

$$R(x) = \max\{\Psi(x, u^*), \Phi(x), R_A\}. \qquad (3.5)$$

The character of the equilibrium urban configuration is crucially affected by how accessibility measure $A(x)$ is specified. Let us consider two examples:

$$A(x) = \int b(y) e^{-\alpha|x-y|}\, dy, \qquad (3.6)$$

$$A(x) = \int b(y)(a - \alpha\,|x - y|)\, dy, \qquad (3.7)$$

where $b(y)$ is the number of business firms per distance at y. When $A(x)$ is defined by (3.6), we call it a *spatially-discounted accessibility measure*; when defined by (3.7), a *linear accessibility measure.* For the latter case of (3.7), Ogawa and Fujita [99] and Imai [67] show that an equilibrium urban configuration uniquely exists under each set of parameters, and that each equilibrium configuration belongs to one of three categories, *monocentric configuration, completely mixed configuration,* or *incompletely mixed configuration.* These configurations are depicted in Figure 3.1–3.3, where RD represents an (exclusive) *residential district,* BD an (exclusive) *business district,* and ID an *integrated district* (i.e., mixed district). Thus, subcenters cannot arise in the case of linear accessibility measure. The more interesting case is with a spatially discounted accessibility measure.

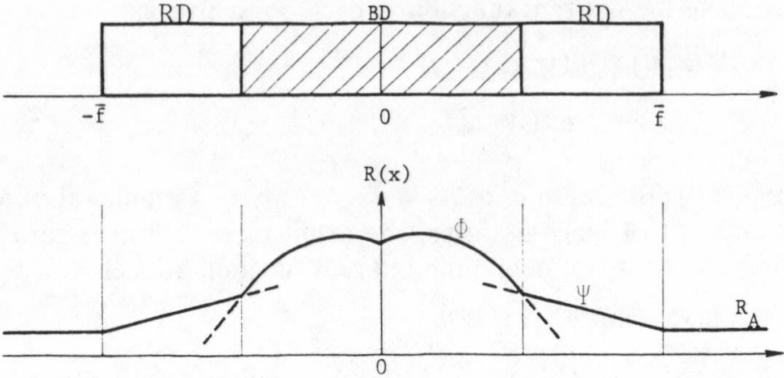

FIGURE 3.1 Monocentric configuration.

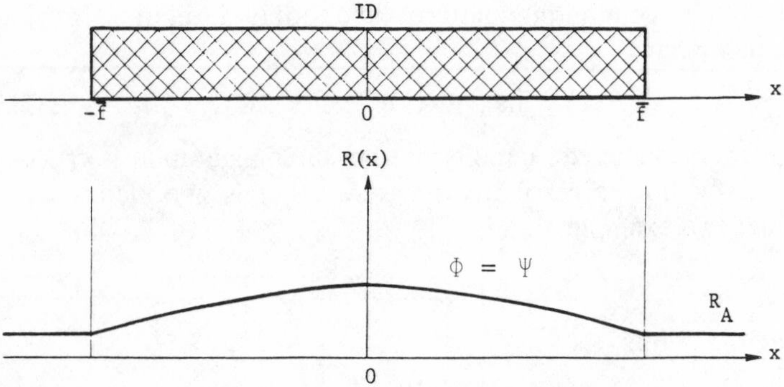

FIGURE 3.2 Completely mixed configuration.

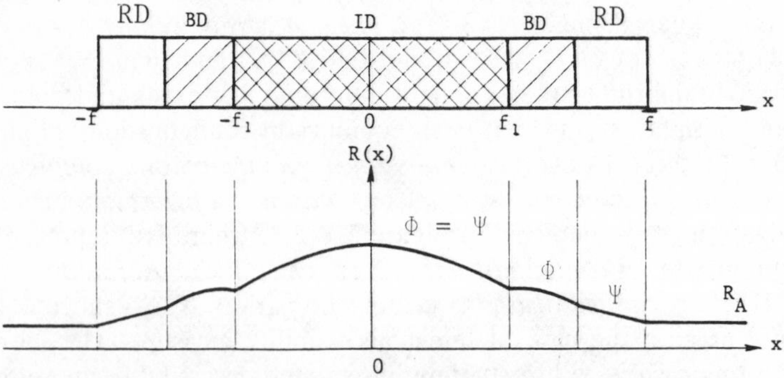

FIGURE 3.3 Incompletely mixed configuration.

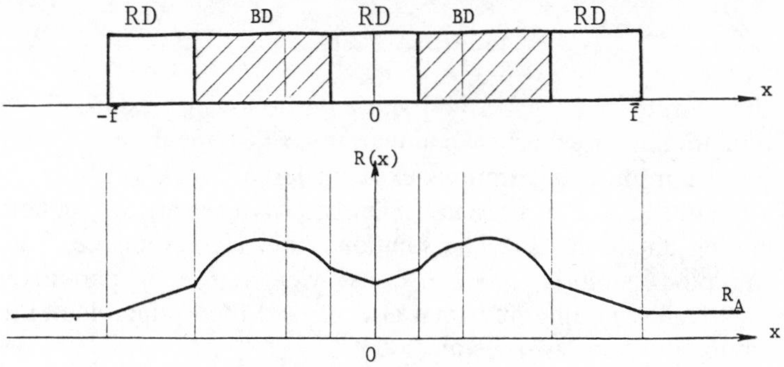

FIGURE 3.4 Duocentric configuration.

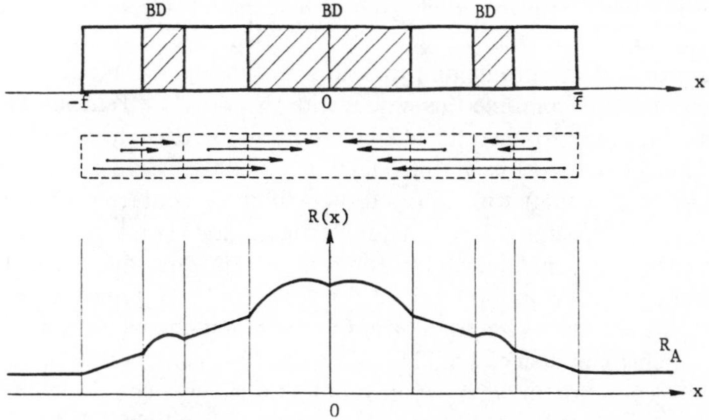

FIGURE 3.5 Tricentric configuration of type *A*.

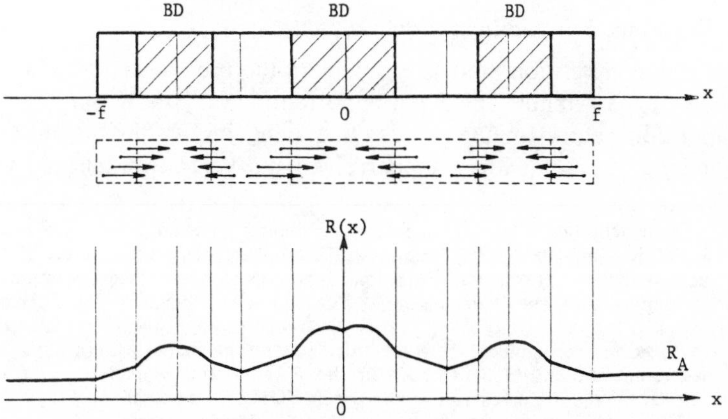

FIGURE 3.6 Tricentric configuration of type *B*.

For this, Fujita and Ogawa [47] show the following results. (i) In addition to the three urban configurations just mentioned, many other equilibrium configurations exist, including ones with subcenters. Figures 3.4–3.6 present examples, where arrows denote commuting directions. (ii) The solution is not always unique. That is, multiple equilibria occur over a wide range of parameter values.[57] Ogawa, Fujita and Imai also studied the optimal land use pattern for the same two-sector model.[58]

A complete general equilibrium model of spatial externalities would simultaneously determine the size, number and contact pattern of firms with the equilibrium urban spatial structure. To this end, a number of improvements are needed. We may use more general utility and production functions in the Ogawa–Fujita–Imai model. We may combine this model with the work by Tauchen and Witte [142, 143] so that the contact pattern of each firm will be determined endogenously. Introduction of different types of firms (and/or households) with differing benefit from contact is also an important extension. Smith and Papageorgiou [133] present a somewhat different, but interesting model in this direction. In addition, the assumption of fixed firm size could be removed. We should also characterize equilibrium and optimal configurations in a two-dimensional space. Finally, by replacing business firms in the above two-sector models with retail stores, we could develop shopping-center models as Papageorgiou and Thisse [106] indicate.

3.4. Imperfect competition model: Type C

Imperfect competition models supplement the previous two types of models (Type A and Type B) in important ways. Briefly, imperfect competition refers to non-price-taking behavior within the urban economy (i.e., behavior with perceived market power effects on

[57] These conclusions were obtained by combining algebraic and numerical analyses. More complete algebraic analysis of the problem is left for the future.

[58] These studies by Ogawa, Fujita and Imai as well as all other studies of optimal land use pattern for Type B models (e.g., Borukhov and Hochman [24], O'Hara [100]) reached the same conclusion. That is, since only positive spatial externalities are considered in these studies, the competitive equilibrium spatial pattern tends to have less concentration of activities than the Pareto-optimum pattern. This is because each agent determines its location by considering the external benefits which it receives from others while ignoring the benefits it bestows on others.

prices and market areas, in particular, for labor and land). For example, due to scale economies or indivisibility, if the number of agents of a type is small compared with the overall urban economy, then these agents may behave monopolistically. Even if the number of agents of a type is large, each agent may retain some monopolistic power in its market area. Since much work has not been done on this subject, we will only enumerate conceivable models here.

C-1. The factory town with single firm: This is the simplest class of Type C model. It involves a city with a single monopolistic firm and workers (e.g., Mills [80], Kanemoto [70, Ch. 2]). Assume that the firm can purchase the entire extent of the city land at the agricultural land price, and that workers can move costlessly across the city boundary and hence the utility of workers equals that of the rest of the economy. Then, the firm will behave as a utility taking developer, and will plan the production activity and the land use of the entire city so as to maximize the total net revenue from production and land development. In the absence of externalities, the optimal residential land use here will coincide with the equilibrium residential land use of the open city model (Section 1.2) under an appropriate wage rate. An important question is left unanswered for this model. That is, although we have assumed *a priori* that the firm can purchase the entire extent of the city land at the agricultural land price, under what condition is this actually possible? In addition to the free mobility of the firm and the competitive agricultural land market, it is necessary to assume that the agricultural land is sufficiently large relative to the land requirement for the city. But, what is the minimum amount of agricultural land required? A related question is, if the firm's profit is affected by the distance to a particular location (e.g., a port or mine), what will be the purchase price of land for the city by the firm?

C-2. The factory town with multiple firms: We may introduce more than one firm into the above factory town model. In order to prevent the uninteresting result that each firm just form its own city independently of others, we may introduce the following alternative assumptions: (i) All firms are identical, but the total land is bounded and small, (ii) all firms are identical and the total land is unbounded, but the profit of each firm is affected by the distance to a common point such as the port or mine, or (iii) firms produce different goods, and trades of goods occur either among firms or between firms and consumers.[59] Eventually, we must determine the equilibrium number of firms and their distribution in the area.

C-3. Spatial competition with land market: For illustration of this, consider a scenario in which two independent supermarkets are planning to locate in a residential area. The location of the supermarkets affects the distribution of households and land rent in the area. The competition for customers will tend to force the two supermarkets towards the center. However, even neglecting the land consumption by each supermarket, high land rent will decrease the purchasing ability of consumers. Thus, the perceived effects on land rent will tend to keep them apart. This is a Hotelling-problem with a land market. Fujita and Thisse [48] reports some interesting results on this problem. A similar problem could be developed in the context of the Lösch model.

C-4. Commercial/Shopping center models: Simultaneously considering both aspects of spatial competition and agglomeration economies, we will be able to develop a variety of models in order to explain the formation of commercial/shopping centers.

C-5. Competition among communities: We can conceive a variety of models of competition among communities which provide local public goods to their residents. This subject is discussed in detail by Henderson, Kanemoto and Schweizer.[60]

We could generate many interesting problems by appropriately fusing different models in the above three categories, A, B and C. It would be wise, however, to thoroughly study each *pure* model first.

4. DYNAMICS

In the previous three sections, we reviewed the static theory of urban land use. Although it has significantly improved our understanding of the city, the static theory of urban land use is intrinsically limited in its usefulness because it completely neglects

[59] A problem similar to the one under assumption (iii) was studied by Schulz and Stahl [127], and a number of surprising results were obtained. That study differs in that the land market is absent, the distribution of workers is exogenously given, and the focus is one the influence of firm location on the wage distribution.

[60] See survey articles by Henderson, Kanemoto and Schweizer in this series.

the durability and adjustment costs of urban infrastructure. Indeed many urban spatial phenomena related to growth can not be explained or evaluated by static models, yet they are quite important to both positive and normative studies of real cities. Among these are urban sprawl, urban decay, urban renewal, the filtering process in the housing market, and effects of uncertainty on land development and land prices.

Dynamic models can address such concerns. With them we may better explain the spatial growth of the city as a historical process, evaluate the efficiency of the process, and suggest means of controlling the process. In this section, we present an overview of the state of the art in dynamic models of urban land use, and suggest future research directions. Though still in their infancy, these models have already revealed that the city is a much more complex system than implied by static models, and that dynamic models can provide more accurate pictures of cities in the real world. It appears that the study of urban spatial dynamics is one of the most promising fields in urban economics today.[61]

4.1. Prototypes of dynamic models

Durability and costly replacement of urban infrastructure are central considerations in all dynamic models of urban land use. This signifies the importance of expectations about the future in determining present decisions about land development. Hence, one way to classify dynamic models of urban land use is based on how market participants perceive the future. To that end, we may first focus on the following three extreme types of models: perfect foresight models, static foresight models, and rational expectation equilibrium models (refer to Figure 4.1).

In *perfect foresight models* (located at Vertex A in Figure 4.1), all market participants are assumed to have perfectly accurate foresight about future market variables such as land prices and housing rents; we study the character of spatial growth paths for which expectations are always realized. An opposing starting point is vertex B (*static foresight models*) where all market participants are assumed

[61] For more detailed discussion on the contents of this section, refer also to Fujita [42].

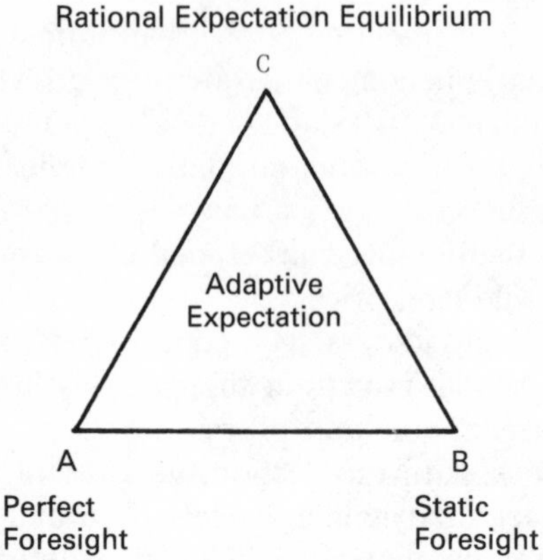

FIGURE 4.1 Conceptual framework for modeling expectation behaviors (modified version of Anas [4]).

to be completely myopic and to base their decisions on currently known prices and other variables. Since all the models on segment AB take the form of point expections, thus no element of uncertainty is explicitly considered. When uncertainty is considered, models must specify both the expectation behaviors of market participants and the particular manner in which these participants weigh alternative uncertain prospects. One good starting point for constructing the theory of dynamic urban land use with the uncertainty dimension is vertex C (*rational expectation equilibrium models*), where market participants have stochastic expectations (about all future variables) under which plans made by all participants are mutually consistent at each time under every conceivable event in the future. Theoretically, this is a logical extension of perfect foresight models to the uncertain world.

Once we leave the corners of the triangle of Figure 4.1, we move into the world of *adaptive models* where processes of expectation formation and learning by market participants as well as the market adjustment mechanism are specified in a variety of ways.

Most current theoretical studies on urban spatial dynamics belong to one of the three vertices of Figure 4.1. This may reflect the time-honored practice of theory building in economics in which conceptually simple extreme models are examined first. For many reasons, this section focuses mostly on perfect foresight models.

Pedagogically, it is preferable to study the pure interaction between space and time prior to examining the effects of uncertainty; and currently there are few studies of urban spatial growth under uncertainty. Static foresight models were advocated, among others, by Anas [5, 6], Brueckner [26, 27] and Harrison and Kain [52]. Although models with static expectations are easy to solve, they have major drawbacks. For example, according to the assumption of static foresights, development occurs over time in successive rings from the CBD outwards, so urban sprawl never occurs. Wheaton notes: "A central criticism of the Anas, or Harrison and Kain approach would seem to be that urban development rarely occurs with myopic foresight. While few observers would contend that truly long-run changes are anticipated, present value calculations substantially diminish the importance of such knowledge anyway. More realistically, urban development does occur under prior estimates about the 'relevant' future. The existence of an active, speculative market in urban land suggests that the anticipation of future events is fully part of the dynamic process" (Wheaton [151, p. 2]). Another important justification for the study of perfect foresight models comes from a normative reason. Although solution paths of a perfect foresight model may not always be efficient, their study can often suggest how to achieve an efficient spatial growth through appropriate market controls.

It is helpful to compare some fundamental differences between the land market in static models and that in dynamic models. In static models, equilibrium land allocations are achieved through the mediation of land rent. Therefore, the concept of *bid land rents* plays a key role. On the other hand, a basic assumption of dynamic models is that land is serviceable for urban use only when buildings or other structure are constructed on it so that the adjustment of land use is very costly. Therefore, except for agricultural use, the concept of land rent cannot be well defined; and the equality between demand and supply of land at each location at each time is achieved through the mediation of land price (i.e., asset price of land), instead of land rent. Thus, the concept of *bid land prices* plays the central role in the dynamic theory. Next, in static models, it is considered that bid land rents are bid by renters of land. In dynamic models, however, the rental market for land cannot be well defined (except for agricultural use). Although a rental market will exist for buildings, renters of buildings may change overtime. Moreover, many of future users of land and buildings may not yet exist in the land market today. Therefore, it is not appropriate to consider that bid land prices are bid by renters of land or buildings. In dynamic models, it is convenient to introduce a class of agents called *land developers,* and to consider that bid land prices are bid by them. The role of developers is to anticipate the demand for land in the future, to plan how to use each piece of land, and to be responsible for land-conversions at each time. Assuming that transactions of real assets are costless, we can even consider that developers own the land and buildings. Hence, they are the main actors in dynamic models.

4.2. One sector model

A simple class of dynamic residential growth models under perfect foresight has been developed by Anas [7], Arnott [13], Fujita [41] and Wheaton [151]. All the studies address the same question using essentially the same framework. That is, they study the equilibrium growth process of residential land, assuming a monocentric city with identical households, where all developers have perfect foresight about the future time path of housing rent. Housing is assumed to be infinitely durable with no demolition considered. Ohls and Pines [101] and Pines [108] studied the same problem in normative context. We present here a summary of these studies following mainly Fujita [41] and Wheaton [151].

Assume, for simplictly, that each residential house is characterized by only its lot size, s.[62] Suppose that all developers are at time 0 (now). Then, each feasible *land use plan* can be represented by (s, t), where t is the time to convert agricultural land to houses of lot size s. We denote by r the distance from the CBD, and assume also that no *tulip-mania expectation* prevails in the land market.[63]

[62] As in Wheaton [151], we may alternatively assume that each house is characterized by lot size s and the amount of housing capital k. However, the inclusion of k does not affect the analysis or conclusions substantially.

[63] We assume that $\lim_{\tau\to\infty} e^{-\gamma\tau}P(r, \tau) = 0$ and $\lim_{\tau\to\infty} e^{-\gamma\tau}P_H(r, s, \tau) = 0$, where $P(r, \tau)$ is the land price at r at time τ and $P_H(r, s, \tau)$ is the price of a house, (r, s), at time τ. This assumption prevents the possibility of tulip-mania expectations, and guarantees the efficiency of the competitive equilibrium path under perfect foresight. See Fujita [42, section 3.3] and footnote 65.

Then, the present value of the stream of net revenues from a unit of land at r when plan (s, t) is adopted there, or the *bid land price under plan* (s, t) *at location r at time* 0, can be calculated as follows:

$$\hat{P}(r, 0 \mid s, t) = \int_0^t e^{-\gamma\tau} R_A(\tau)\, d\tau + \frac{\int_t^\infty e^{-\gamma\tau} R(r, s, \tau)\, d\tau - e^{-\gamma t} C(s, t)}{s}, \tag{4.1}$$

where γ is the time discount rate or interest rate, $R_A(\tau)$ the agricultural land rent at time τ, $R(r, s, \tau)$ the rent for a house with lot size s at r at time τ, and $C(s, t)$ the construction cost of a house with lot size s at time t. All of γ, $R_A(\tau)$ and $C(s, t)$ are assumed to be exogenously given.

Let us denote by $P(r, t)$ the price of a unit of land at distance r at time t. Then, we can immediately see that on the competitive equilibrium path, the following relations must hold:

Rule 4.1. $P(r, 0) = \max_{s,t} \hat{P}(r, 0 \mid s, t)$: The initial land price curve is the upper envelope of all feasible bid-price curves.
Rule 4.2. A plan (s, t) will be chosen at location r only if $\hat{P}(r, 0 \mid s, t) = P(r, 0)$, that is, only if the bid land price under plan (s, t) is maximum among all feasible bid land prices there.
Rule 4.3. If some vacant land remains at location r at time t, $(\dot{P}(r, t) + R_A(t))/P(r, t) = \gamma$, where "·" denotes the time derivative. That is, the sum of capital gain and income gain per unit of asset price equals the interest rate (*arbitrage condition*).

We may call (s, t) an optimal plan at location r if $\hat{P}(r, 0 \mid s, t) = P(r, 0)$. For example, if (s^*, t^*) is an optimal plan at location r^*, then from Rule 4.1, the relationship between the initial land price curve $P(r, 0)$ and bid land price curve $\hat{P}(r, 0 \mid s^*, t^*)$ can be depicted as in Figure 4.2.

Next, in order to determine the equilibrium house rent $R(r, s, \tau)$, we assume that all residential houses are owned by outside developers and rented to households. Each household rents one house at each time, and can move among rental houses without cost. The households have identical utility function $U(z, s)$ and the same income $y(\tau)$ at each time τ, which is spent on the composite consumer good z, a house and transport costs. Then, the equi-

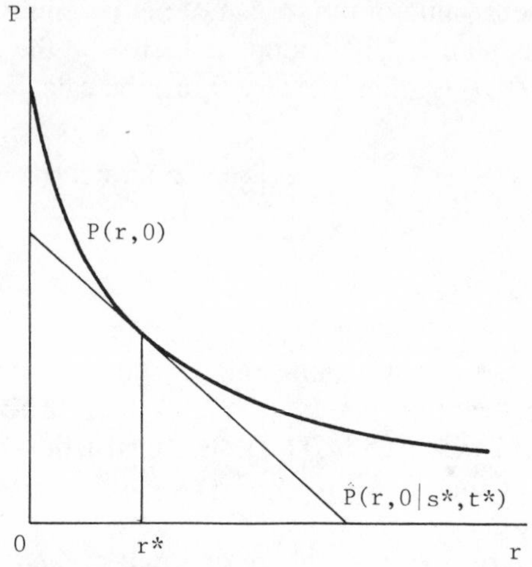

FIGURE 4.2 Initial land price curve $P(r, 0)$ and bid land price curve $\hat{P}(r, 0 \mid s^*, t^*)$.

librium house rent function is obtained as follows:

$$R(r, s, \tau) = y(\tau) - T(r, \tau) - Z(s, u(\tau)), \tag{4.2}$$

where $T(r, \tau)$ is the commuting cost function, and $Z(s, u(\tau))$ is the solution of $U(z, s) = u(\tau)$ for z.

In the *open city model,* utility level $u(\tau)$ at each time τ is exogenously specified, to coincide with the utility level of the national economy at that time. Unknowns are the equilibrium population stream $N(\tau)(0 \leqq \tau < \infty)$ and the equilibrium residential development process which meets $N(\tau)$. On the other hand, in the *closed city model,* population stream $N(\tau)$ is exogenously given, but $u(\tau)$, $R(r, s, \tau)$ and residential development process are unknowns. In either case, all unknowns can be determined by using Rules 4.1 to 4.3 and relation (4.2).

For either open or closed city models, the major concerns are:

A) Whether the development of land occurs from the CBD outward, or from the edge of the city inward.

B) Whether housing lot size increases with distance from the CBD, or decreases (at least in certain areas).

Let $r(t)$ be the location at which development of land occurs at time

t, and $s(t)$ be the lot size of houses constructed at location $r(t)$ at time. Fujita [41] and Wheaton [151] show that the following five patterns, (a)–(e), are theoretically possible depending on the value of parameters in the problem:

$$\text{(a) } \dot{r}(t)>0,\ \dot{s}(t)>0,\ \text{(b) } \dot{r}(t)>0,\ \dot{s}(t)<0,\ \text{(c) } \dot{r}(t)>0,\ \dot{s}(t)=0$$

$$\text{(d) } \dot{r}(t)<0,\ \dot{s}(t)<0,\ \text{(e) } \dot{r}(t)=\pm\infty,\ \dot{s}(t)=\pm\infty \text{ and } \dot{r}(t)\dot{s}(t)>0,$$

where "·" denotes the derivative with respect to time t. Note that an increase (decrease) in s means a decrease (increase) in population density. Hence, pattern (a) means that the development of land occurs from the CBD outward, and population density decreases with distance from the CBD. This pattern tends to happen when the growth rate of income continues to be sufficiently large compared with the growth rates of the population and transport cost in the city. This pattern was observed in most of the U.S. cities during the steady economic growth period between the end of World War II and the oil price hike in 1973. Pattern (b) means that the development occurs from the CBD outward, and density decreases in distance from the CBD. This pattern tends to happen when the growth rate of population continues to be sufficiently large compared with the growth rate of income and the rate of decrease in transport cost. This pattern can be observed today in many fast growing cities in developing countries. Pattern (c) represents the boundary case between patterns (a) and (b). The most surprising case is pattern (d), which means that the development occurs from the outside inward, and density decreases with distance from the CBD. Note that the outside-inward development represents a kind of leap-frog development. Although this pattern is rarely observable in actual cities, it is theoretically possible when the price elasticity of compensated demand for land is very high and/or transport cost is rapidly increasing with time. Finally, pattern (e) means that the entire land is developed instantaneously (from the inside outward, or from the outside inward), and density decreases with the distance from the CBD. Note that this case of instantaneous completion of the whole city is mathematically equivalent to the basic model of Section 2. That is, the static model of Section 2 can be considered as a special case of the dynamic model here.

In short, the one sector model of residential growth is useful in demonstrating how land development will be affected by expectations about the future. Many characteristics of its solution are interesting and often quite different from those predicted by static models or by static foresight models. For example, population density may increase with commuting distance in certain areas; this phenomenon never occurs in static models. For another example, land development may occur from the outside inward; this never happens in static-foresight models.[64]

Still, the one sector dynamic model has limitations. In particular, the residential area always expands densely without leaving vacant land inside. This is due to the assumption of single sector. In order to study the phenomena of urban sprawl more generally, we need a model with more than one activity type. This is the subject of the next subsection.

4.3. Urban sprawl

Urban sprawl has been one of the most controversial issues in urban economics and city planning, yet is it not satisfactorily understood theoretically. Clawson [32] defines sprawl as "the lack of continuity in expansion." In the context of monocentric city models, Mills [79] cites three land use patterns that qualify as examples of sprawl: (A) *Leap-frog development,* when a von Thünen ring of undeveloped land separates rings of developed land. This form of sprawl involves radial discontinuity. (B) *Scattered development,* when there are annuli with both developed and undeveloped land in them. (C) *Mixed development,* when there are annuli with more than one developed use.

We cannot deny that there are costs associated with sprawl, particularly those costs associated with long commuting distance, public-sector inefficiency and esthetic unattractiveness. It has also been demonstrated in Fujita [42, Section 3.3] that urban sprawl due to tulip-mania expectation generally leads to inefficiency.[65] How-

[64] For details, see Fujita [41] and Wheaton [151] as well as Anas [7], Arnott [13], Ohls and Pines [101] and Pines [108].

[65] For the "tulip-mania phenomenon" in durable asset markets, see Samuelson [121, 122]. In short, a tulip-mania phenomenon occurs when people happen to hold very strong expectations of future land prices which cannot be fully explained by the real-value of land (i.e., present value of net revenues from actual utilization of land). The realism of tulip-mania will increase when uncertainties in future urban growth are introduced, but the basic causes of tulip-mania phenomena are the same; land is an infinitely durable good, and there is no end to the future.

ever, many authors have recently suggested that in a growing city, efficiency may require sprawl-fashioned development. Fujita [41], Ohls and Pines [101], Pines [108] and Wheaton [151] have shown that leap-frog development often occurs in an efficient land development process in one-sector model. Multisector models (i.e. with more than one type of building-using activity) are needed to fully investigate sprawl-phenomena. Thus far, studies by Fujita [37, 38, 39] with a fixed-coefficient model and a numerical study of Pines and Werczberger [109] with a linear programming model indeed suggest that when the growth period of the city is sufficiently long, the efficient land development for a multi-sector model generally exhibits all the three types of sprawl-phenomena mentioned previously. In the following we present a simple two-sector model of land development under perfect foresight, and explain the mechanism of urban sprawl due to a long-run efficiency requirement.

Suppose that there are two building-using activities ($i = 1, 2$) in the city, and that each unit of type i activity uses one and only one unit of type i building at each time. We may consider, for example, that $i = 1$ represents firm-buildings, and $i = 2$ residential houses. For simplicity, let us assume that each unit of building i occupies a fixed lot size, s_i ($i = 1, 2$). As in the previous subsection, we assume a monocentric city, and denote by r the distance from the city center. It is also assumed that each building is infinitely durable with no demolition considered. We denote each feasible *land use plan* by (i, t) where t is the time to convert agricultural land to buildings of type i ($i = 1, 2$, $t \geqq 0$). Then, similarly to equation (4.1), the bid land price under plan (i, t) at location r at time 0 can be defined as follows:

$$\hat{P}(r, 0 \mid i, t) = \int_0^t e^{-\gamma\tau} R_A(\tau)\, d\tau + \frac{\displaystyle\int_t^\infty e^{-\gamma\tau} R_i(r, \tau)\, d\tau - e^{-\gamma t} C_i(t)}{s_i}, \tag{4.3}$$

where $R_i(r, \tau)$ is the rent for a unit of building i at r at time τ, and $C_i(t)$ the construction cost per unit of building i at time t. For simplicity, let building rent $R_i(r, \tau)$ be linearly decreasing in distance r:

$$R_i(r, \tau) = R_i(0, \tau) - a_i(\tau) r, \quad i = 1, 2, \tag{4.4}$$

where $a_i(\tau)$ is a given, positive, continuous function of time τ.[66] Then, from (4.3),

$$\partial\hat{P}(r, 0 \mid i, t)/\partial r = -A_i(t)/s_i, \quad \text{where} \quad A_i(t) = \int_t^{\infty} e^{-\gamma\tau} a_i(\tau)\, d\tau. \tag{4.5}$$

Let us assume that $a_1(\tau)/s_1 > a_2(\tau)/s_2$ for all τ, and hence

$$A_1(t)/s_1 > A_2(t)/s_2 \quad \text{for all} \quad t \geqq 0. \tag{4.6}$$

That is, the bid land price curve for building type 1 is always steeper than that for building type 2.

Next, we examine the equilibrium process of land development assuming that

$$\dot{N}_i(t) > 0 \quad \text{for} \quad t \in (0, \bar{t}), \qquad = 0 \quad \text{for} \quad t \geqq \bar{t}, \quad i = 1, 2. \tag{4.7}$$

That is, the city grows from time 0 to time $\bar{t}$, and thereafter stays the same.[67] This implies that at each time $t \in (0, \bar{t})$, $\dot{N}_i(t)$ units of building i must be constructed at some distance ($i = 1, 2$). Applying Rule 4.2 in the present context, we can see that the construction of type i buildings at each time t occurs at distance $r_i(t)$ where the initial land price curve $P(r, 0)$ and bid land price curve $\hat{P}(r, 0 \mid i, t)$ are tangent (see Figure 4.3). Therefore, from (4.6), we can conclude that construction of building type i always occurs closer to the city center than that of building type 2:

$$r_1(t) < r_2(t) \quad \text{for all} \quad t \in (0, \bar{t}). \tag{4.8}$$

And, since slope $A_i(t)/s_i$ of each bid land price curve is decreasing in t, we can also conclude that construction distance of each building type continuously moves outward with time:

$$\dot{r}_i(t) > 0 \quad \text{for} \quad t \in (0, \bar{t}), \quad i = 1, 2. \tag{4.9}$$

Assuming that $\bar{t} < \infty$, there are essentially two different cases for

[66] With $i = 1$ representing firms and $i = 2$ households, these linear building rent functions are based on the following two implicit assumptions: (i) the transport cost function for firms' outputs to the trade node at the city center is linear in distance, and (ii) households always commute inwardly and the commuting cost function is linear in distance.

[67] For other examples of land development processes, and for the calculation procedure for the solution path, see Fujita [38, 39, 42].

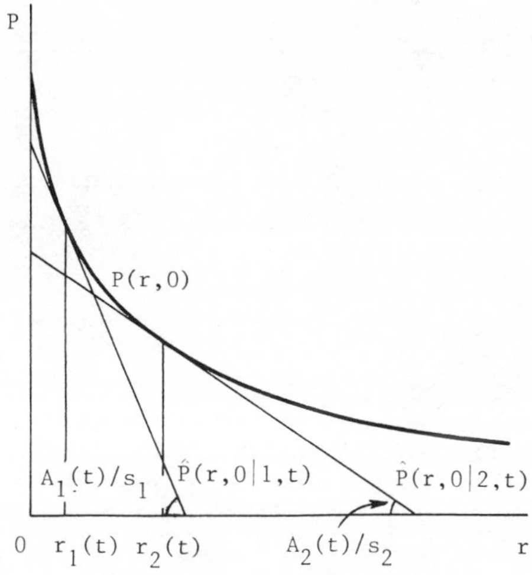

FIGURE 4.3 Determination of construction distances $r_i(t)$ $(i = 1, 2)$ at each time t.

the pattern of development:

$$\text{case } a\text{: } A_1(\bar{t})/s_1 < A_2(0)/s_2, \qquad \text{case } b\text{: } A_1(\bar{t})/s_1 \geqq A_2(0)/s_2.$$

Case a occurs when growth period $\bar{t}$ is sufficiently long. From (4.6), $r_1(0) = 0 < r_2(0) < r_1(\bar{t}) < r_2(\bar{t})$. Therefore, recalling (4.9), we can summarize the equilibrium process of land development in this case as in Figure 4.4. The height of each diagram represents the ratio of land occupied by each building at each distance. We can see from the figure that this process of land development exhibits all the three types of urban sprawl: leap-frog development (i.e., type 2 buildings are initially constructed at some distances from the areas where type 1 buildings are being constructed), scattered development (i.e. construction of type 2 buildings does not exhaust all available land at each distance of construction), and mixed development (i.e., buildings of both types eventually mix in large area). We can also prove that when it occurs, this equilibrium process of sprawl-fashioned development is also socially most efficient (see Fujita [37, 38]). The land development process under case b is simpler. Here, $r_1(0) = 0 < r_1(\bar{t}) = r_2(0) < r_2(\bar{t})$. Therefore, construction of type 1 buildings begins from the city center, and construction of type 2 buildings begins at some distance far from the city center.

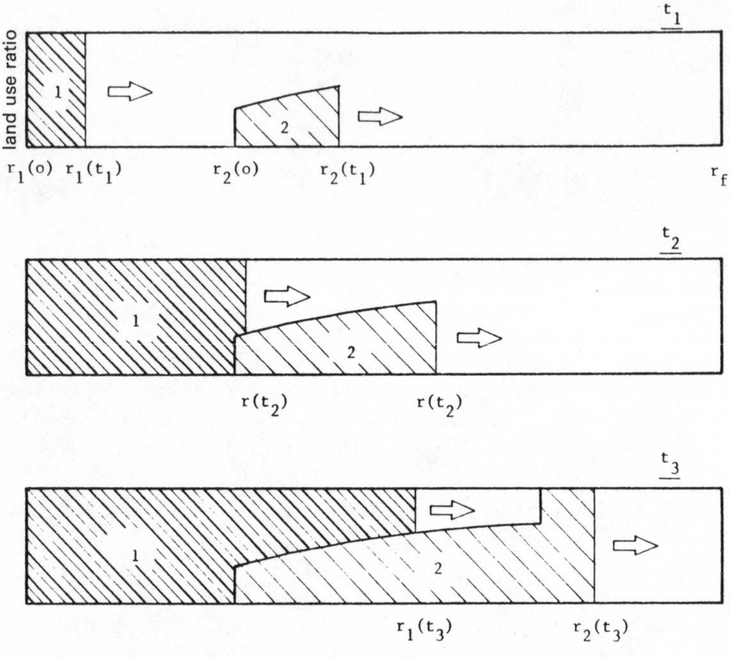

FIGURE 4.4 Land development process: $n = 2$, $(t_1 < t_2 < t_3 < \bar{t})$.

And, at the end of growth period, the city consists of two Thünen rings. This process of land development exhibit only one type of sprawl-phenomena (i.e., leap-frog development), and this pattern occurs only when the growth period of the city is very short.

The intention of our analysis is not to maintain that urban sprawl always reflects efficient market processes. Rather, it suggests that "competitive urban land markets may well allocate land more efficiently than many observers believe" (Ohls and Pines, [101, p. 234]). If this is the case, we must be very careful in regulating the land market so as not to discourage the positive contributions of land speculation.

4.4. Urban renewal, filtering process, and land development under uncertainty

In this subsection, we briefly summarize the state of the art on three other main issues of urban spatial dynamics: urban renewal, the filtering process in housing market, and urban land development under uncertainty. Here, we are concerned with only those studies which explicitly treat 3 dimensions—space, time and durability.[68] Many other works related to these three issues do not explicitly consider either spatial dimension or durability of building.

A. Land development with renewal. So far, we have neglected the issues of building maintenance and renewal. Although these are important issues in actual land development, they make the study of the urban land development process analytically complex. With perfect foresight models, the problem is intractable until some simplifying assumptions are introduced. In the case of static foresight models, the solution is much easier to obtain. We shall summarize each case briefly.

Akita and Fujita [1] introduced the possibility of renewal into the fixed-coefficient model of the previous subsection, and studied the land development process under perfect foresight. This model demonstrates that at the initial stage, land is developed in a sprawl-fashioned manner as depicted in Figure 4.4; then, in a later stage, renewal of buildings from type 2 to type 1 starts from the area close to the city center. Fujita [37] proposed a more general model of urban spatial growth with possibility of land renewal. Since deterioration of buildings is not considered in these models, renewals in these studies is due solely to the efficiency of the spatial allocation of different building stocks over time. On the other hand, Brueckner [28] studies a one-sector model of residential renewal in a steady-state environment where dwellings deteriorate as they age. It is shown that under the assumption of zero demolition costs, the optimal land use plan at each location consists of an infinite sequence of identical buildings. Pines and Werczberger [109] developed a linear programming model of land development with renewal, and presented a numerical example for discontinuous sprawl and renewal. Brueckner [26] studied a vintage model of urban growth with static foresight by developers. Wheaton [152] studied a similar problem without consideration of building deterioration. With the aid of computer simulations, both studies

[68] For additional works related to the above three issues, see Fujita [42]. Also, for a more detailed review of housing market, see the survey article by Muth and Quigley in this series.

reproduced the phenomenon of "Manhattanization," where part of an aging central city was replaced by new skyscrapers which dwarf older surrounding structures. Brueckner [27] performed a similar study with two income classes, and demonstrated that spatial mixing of the two income groups might occur. Although these static foresight models have the advantage of tractability and often reproduce important spatial growth phenomena, they can provide little information about efficiency.

All the dynamic models mentioned so far explicitly consider that land is serviceable for urban use only when structures are built on it so that the adjustment of land use is very costly. Hochman and Pines [65, 66] consider this premise less explicitly, and developed a different class of one-sector residential development model. They assume that the rate of net increase, $\dot{H}_j(t)$, in housing stock $H_j(t)$ in zone j at time t is

$$\dot{H}_j(t) = h(H_j(t), C_j(t), L_j), \tag{4.10}$$

where $C_j(t)$ represents the capital input, and L_j the area of zone j. Since it is assumed that both $\partial h/\partial H$ and $\partial^2 h/\partial H^2$ are negative ("reflecting the increasing difficulty of adding new units of housing as density increases"), the amount of housing stock in each zone can increase, decrease, or stay constant. In the context of an open city model, they performed detailed analyses of land development, mostly under the assumption of perfect foresight. Although the model is analytically more tractable and many results are similar to those in Fujita [41] and Wheaton [151], there are some essential differences. In particular, for Hochman and Pines [65, 66] density always decline with distance from the city center as well as housing rent and housing values, and no phenomenon of urban sprawl occurs with optimal development. This is because housing stock is assumed to be completely malleable within each zone.

B. Filtering process in the housing market. In this process, dwellings are bid away from their original occupants. To date, most studies of the filtering process are either aggregate models (i.e., without spatial dimension) or partial models (i.e., short-run dynamics of housing market). There are a few exceptions. A vintage model of residential growth (under static foresight) with two income classes was proposed by Brueckner [27], and a number of interesting results were obtained with the aid of computer simulations. Fujita [37] proposed a general model of urban spatial growth (under perfect foresight) with the possibility of housing filtering. While only general characteristics of optimal and equilibrium growth paths were studied there, Ando and Fujita [11] presented a detailed study of growth in the case of two income classes. The latter study demonstrates the difficulties in obtaining concrete results for such a complex problem using a purely analytical method. Although Fujita [37] formulated the problem in continuous time and discrete space, Pines and Werczberger [109] formulated a similar problem as a linear programming problem with discrete time and space; and they presented a numerical example in the context of the open city model. This suggests that computer simulation will be useful in studying complex problems of urban land development.

C. Land development under uncertainty. Thus far, we have discussed only models with point expectations; uncertainly has not been explicitly considered. The future is, however, essentially uncertain and so expectations should generally take the form of stochastic processes. Accordingly, we should specify the following elements for each actor in the land market: (i) probabilistic belief of the actor about the future states of the world, (ii) information structure and learning process of the actor, and (iii) objective function of the actor at each decision time. In trying to develop a theory of urban spatial growth under uncertainty, it would be natural to begin by studying the character of the *rational expectation equilibrium path* (*REE path*) in the sense of Hicks [63] and Radner [111]. At present, two papers, Fujita [40] and Mills [79], address the REE path of urban land development. By using a continuous time framework with a very general stochastic process, Fujita [40] obtained the explicit solution of the REE path for a simple one-sector model of residential development as a function of exogenously given stochastic population and transport conditions in the city. The study shows that since any land with a positive probability of being developed in the future has a land price greater than the agricultural land price, the total area of land with positive *urban* land price tends to be very large at each time. Mills [79] studied a two-period, two-sector model of land development under uncertainty. The study shows that urban sprawl becomes stronger

	Monocentric without Externalities	Monocentric with Externalities	Monmonocentric
Perfect foresight	1		
static foresight	2		
Rational expectation equilibrium	3		
Adaptive			

FIGURE 4.5 Classification of conceivable models of urban spatial dynamics.

with uncertainty, and that *ex post* inefficiency in land development will be generally observed. Since both studies are based on a number of simplifying assumptions, a more complete understanding of urban land development under uncertainty (including the study of non-REE paths) is left for the future.

4.5. Suggestions for further research

Although the models discussed in the previous subsections may help our understanding of various aspects of urban spatial dynamics, they nevertheless represent only a small portion of many conceivable models. This point can be confirmed from Figure 4.5, which presents a classification of conceivable economic models of urban spatial dynamics according to the type of expectation behavior and the degree of generality as an urban spatial model. As we have seen, most existing models belong to either box 1 or box 2 in this figure, a few to box 3, and essentially none in other boxes. In the following, we briefly discuss the possible extensions of existing models in the framework of the first column. Eventually we must move towards the second and third columns, but that is beyond the scope of this discussion.[69]

[69] See Fujita [42] for further discussion on future research, including possible extensions in the frameworks of the second and third columns.

(i) *extension of perfect foresight models.* First, we may introduce multiple income classes into the model discussed in subsection 4.2. Then, we can study both urban sprawl and the housing filtering process in a single model. Ando and Fujita [11] report initial results from this extension. Second, we may extend the models of 4.4.A to further advance the study of urban renewal process. Of particular interest is whether (or under what conditions) the spatial pattern of the city gets closer to the pattern of the static Alonso–Muth model as renewals are repeated. Third, we may develop a residential growth model based on long-run utility maximization by households; in particular, a model based on permanent income will be promising. Then, we can investigate a number of important distributional issues related to housing/land ownership and urban growth.

(ii) *advancement of numerical solution methods.* Although the formulation of each problem suggested above would be relatively easy, analytic solution may generally be impossible. Thus, research on efficient numerical methods for solving these complex problems becomes important.

(iii) *further research on REE path under uncertainty.* At present there are few studies on land development under uncertainty. First, we may thoroughly study REE paths of land development in the case where all developers are risk-neutral and have identical expectation. For this, the market problem can be reformulated as an optimization problem. Ellson and Roberts [34] represents an initial attempt in this direction. The second step is to study the effects of risk-aversion on land development. For this, we may first generalize the model of Fujita [40] into the case of multiple building types. We also need to develop a measure of efficiency for the evaluation of land development under uncertainty.

(iv) *adaptive models.* So far we have been concerned with only those models which are based on expectation behaviors at the corners of the triangle in Figure 4.1. These expectation hypotheses, however, are not very realistic. Eventually we must move away from the corners of the triangle and develop *adaptive models* of urban spatial dynamics, where processes of expectation formation and learning by market participants as well as the market adjustment mechanism are specified more realistically. A major difficulty in the study of adaptive models is that once we leave the corners of the triangles in Figure 4.1, there is continuum of different types of adaptive models but we do not have *a priori* reasons for choosing a particular model. In this circumstance, the first work may be to develop an appropriate classification scheme of all adaptive models. Next, we may establish a set of criteria for choosing a class of adaptive models with desirable properties. For this purpose, axiomatic approaches may be useful. Empirical tests of theoretical models are also of urgent necessity.

References

[1] Akita, T. and M. Fujita, "Spatial Development Processes with Renewal in a Growing City," *Environment and Planning A,* **14** (1982), 205–223.

[2] Alonso, W., *Location and Land Use,* Cambridge, MA.: Harvard University Press, 1964.

[3] Altmann, J. L. and J. S. DeSalvo, "Tests and Extensions of the Mills–Muth Simulation Model of Urban Residential Land Use," *Journal of Regional Science,* **21** (1981).

[4] Anas, A., "Comments on M. Fujita's 'Optimum Spatial Patterns of Urban Growth'," Paper presented at the 22nd Annual RSA meetings, 1975, Cambridge, Mass.

[5] Anas, A., "Short-Run Dynamics in the Spatial Housing Market," in G. J. Papageorgiou, ed., *Mathematical Land Use Theory,* Lexington, Mass.: Lexington Books, 1976.

[6] Anas, A., "Dynamics of Urban Residential Growth," *Journal of Urban Economics,* **5** (1978), 66–87.

[7] Anas, A., "The Pattern and Timing of Land Development in Long Run Equilibrium Urban Land Use Model," *Geographia Polonica,* **42** (1979), 91–109.

[8] Anas, A. and D. S. Dendrinos, "The New Urban Economics: A Brief Survey," in G. J. Papageorgiou, ed., *Mathematical Land Use Theory,* Lexington, Mass.: Lexington Books, 1976.

[9] Anas, A. and L. M. Moses, "Mode Choice, Transport Structure and Urban Land Use," *Journal of Urban Economics,* **6** (1979), 228–246.

[10] Ando, A., *Development of a Unified Theory of Urban Land Use,* Ph.D. dissertation, University of Pennsylvania, Philadelphia, 1981.

[11] Ando, A. and M. Fujita, "Dynamics of Residential Development with Multiple Income Classes," *Working Papers in Regional Science and Transportation,* No. 19, University of Pennsylvania, Philadelphia, 1979.

[12] Arnott, R., "Optimal City Size in a Spatial Economy," *Journal of Urban Economics,* **6** (1979), 65–89.

[13] Arnott, R. J., "A Simple Urban Growth Model with Durable Housing," *Regional Science and Urban Economics,* **10** (1980), 53–76.

[14] Arnott, R. J., J. G. MacKinnon and W. C. Wheaton, "The Welfare Implications of Spatial Interdependence," *Journal of Urban Economics,* **5** (1978), 131–136.

[15] Arnott, R. J. and J. G. Riley, "Asymmetrical Production Possibilities, the Social Gains from Inequality and the Optimum Town," *Scandinavian Journal of Economics,* **79** (1977), 301–311.

[16] Arnott, R. J. and J. E. Stiglitz, "Aggregate Land Rents, Expenditure on

Public Goods, and Optimal City Size," *Quarterly Journal of Economics,* **93** (1979), 471–500.

[17] Bailey, M. J., "Note on the Economics of Residential Zoning and Urban Renewal," *Land Economics,* **35** (1959), 288–292.

[18] Barr, J. L., "City Size, Land Rent and the Supply of Public Goods," *Regional and Urban Economics,* **2** (1972), 67–103.

[19] Beckmann, M. J., "On the Distribution of Rent and Residential Density in Cities," Paper presented at the Inter-Departmental Seminar on Mathematical Applications in the Social Sciences, Yale University (1957).

[20] Beckmann, M. J., "On the Distribution of Urban Rent and Residential Density," *Journal of Economic Theory,* **1** (1969), 60–67.

[21] Beckmann, M. J., "Equilibrium Models of Residential Land Use," *Regional and Urban Economics,* **3** (1973), 361–368.

[22] Beckmann, M. J., "Spatial Equilibrium in the Housing Market," *Journal of Urban Economics,* **1** (1974), 99–107.

[23] Beckmann, M. J., "Spatial Equilibrium in the Dispersed City," in G. J. Papageorgiou, ed., *Mathematical Land Use Theory,* Lexington, Mass.: Lexington Books, 1976.

[24] Borukhov, E. and O. Hochman, "Optimum and Market Equilibrium in a Model of a City without a Predetermined Center," *Environment and Planning A,* **9** (1977), 849–856.

[25] Brueckner, J. K., "Spatial Majority Voting Equilibria and the Provision of Public Goods," *Journal of Urban Economics,* **6** (1979), 338–351.

[26] Brueckner, J. K., "A Vintage Model of Urban Growth," *Journal of Urban Economics,* **8** (1980), 389–402.

[27] Brueckner, J. K., "Residential Succession and Land Use Dynamics in a Vintage Model of Urban Housing," *Regional Science and Urban Economics,* **10** (1980), 225–240.

[28] Brueckner, J. K. "A Dynamic Model of Housing Production," *Journal of Urban Economics,* **10** (1981), 1–14.

[29] Capozza, D. R., "Employment-Population Ratios in Urban Areas: A Model of the Urban Land, Labor, and Goods Markets," in G. J. Papageorgiou, ed., *Mathematical Land Use Theory,* Lexington, Mass.: Lexington Books, 1976.

[30] Carliner, G., "Income Elasticity of Housing Demand," *Review of Economics and Statistics,* **55** (1973), 528–532.

[31] Courant, P. N. and J. Yinger, "On Models of Racial Prejudice and Urban Residential Structure," *Journal of Urban Economics,* **4** (1977), 272–291.

[32] Clawson, M., "Urban Sprawl and Speculation in Suburban Land," *Land Economics,* **38** (1962), 99–111.

[33] Dixit, A., "The Optimum Factory Town," *Bell Journal of Economics and Management Science,* **4** (1973), 637–651.

[34] Ellson, R. and B. Roberts, "Residential Land Development Under Uncertainty," *Journal of Regional Science,* **23** (1983), 309–322.

[35] Fales, R. L. and L. N. Moses, "Land Use Theory and the Spatial Structure of the Nineteenth-Century City," *Papers and Proceedings of the Regional Science Association,* **28** (1972), 49–80.

[36] Fisch, O., "Spatial Equilibrium with Local Public Goods: Urban Land Rent, Optimal City Size and the Tiebout Hypothesis," in G. J. Papageorgiou, ed., *Mathematical Land Use Theory,* Lexington, Mass.: Lexington Books, 1976.

[37] Fujita, M., "Toward a Dynamic Theory of Urban Land Use," *Papers of Regional Science Association,* **37** (1976), 133–165.

[38] Fujita, M., "Spatial Patterns of Urban Growth: Optimum and Market," *Journal of Urban Economics,* **3** (1976), 209–241.
[39] Fujita, M., "Spatial Patterns of Urban Growth and Contraction: Problem A," *Geographia Polonica,* **42** (1979), 112–148.
[40] Fujita, M., "Urban Land Market under Uncertainty with Infinite Time Horizon: Part I," *Working Papers in Regional Science and Transportation,* No. 59, University of Pennsylvania, Philadelphia, 1981.
[41] Fujita, M., "Spatial Patterns of Residential Development," *Journal of Urban Economics,* **12** (1982), 22–52.
[42] Fujita, M., "Urban Spatial Dynamics: A Review," *Sistemi Urbani,* **3** (1983), 411–475.
[43] Fujita, M., "Towards General Equilibrium Models of Urban Land Use," *La Revue Economique,* 1, 1985, 135–167.
[44] Fujita, M., "Optimal Location of Public Facilities: Area Dominance Approach", *Regional Science and Urban Economics,* 16, 1986, forthcoming.
[45] Fujita, M., "Existence and Uniqueness of Equilibrium and Optimal Land Use: Boundary Rent Curve Approach", *Regional Science and Urban Economics,* 15, 1985, 295–324.
[46] Fujita, M., *Residential Land Use Theory,* Cambridge University Press, forthcoming (1987).
[47] Fujita, M. and H. Ogawa, "Multiple Equilibria and Structural Transition of Non-Monocentric Urban Configurations," *Regional Science and Urban Economics,* **12** (1982), 161–196.
[48] Fujita, M. and J. F. Thisse, "Spatial Competition Under the Influence of Land Market," CORE Discussion Paper No. 8512, Universite Catholique de Louvain, Belgium, 1984.
[49] Goldstein, G. S. and L. N. Moses, "A Survey of Urban Economics," *Journal of Economic Literature,* **11** (1973), 471–515.
[50] Goldstein, G. S. and L. N. Moses, "Interdependence and the Location of Economic Activity," *Journal of Urban Economics,* **2** (1975), 63–84.
[51] Grieson, R. E. and M. P. Murray, "On the Possibility and Optimality of Positive Rent Gradients," *Journal of Urban Economics,* **9** (1981), 275–285.
[52] Harrison, D. Jr. and J. F. Kain, "Cumulative Urban Growth and Urban Density Functions," *Journal of Urban Economics,* **1** (1974), 61–98.
[53] Hartwick, J. M., "Price Sustainability of Location Assignments," *Journal of Urban Economics,* **1** (1974) 147–160.
[54] Hartwick, J. M., "Agglomeration and the Size and Structure of the CBD," mimeograph, Queen's University, Kingston, 1978.
[55] Hartwick, P. G. and J. M. Hartwick, "Efficient Resource Allocation in a Multinucleated City with Intermediate Goods," *Quarterly Journal of Economics,* **88** (1974), 340–352.
[56] Hartwick, J. M., U. Schweizer and P. Varaiya, "Comparative Statics of a Residential Economy with Several Classes," *Journal of Economic Theory,* **13** (1976), 396–413.
[57] Heffley, D. R., "Efficient Spatial Allocation in the Quadratic Assignment Problem," *Journal of Urban Economics,* **3** (1976), 309–322.
[58] Heffley, D. R., "Competitive Equilibria and the Core of a Spatial Economy," *Journal of Regional Science,* **22** (1982), 423–440.
[59] Helpman, E. and D. Pines, "Optimal Public Investment and Dispersion Policy in a System of Open Cities," *American Economic Review,* **70** (1980), 507–514.
[60] Helpman, E., D. Pines and E. Borukhov, "The Interaction Between Local

Government and Urban Residential Location: Comment," *American Economic Review,* **66** (1976), 961–967.
[61] Henderson, J. V., *Economic Theory and the Cities,* New York: Academic Press, 1977.
[62] Herbert, J. D. and B. H. Stevens, "A Model of the Distribution of Residential Activity in Urban Areas," *Journal of Regional Science,* **2,** No. 2 (1960), 21–36.
[63] Hicks, J., *Value and Capital,* 2nd ed., Oxford: Clarendon Press, 1946.
[64] Hochman, O. and H. Ofek, "The Value of Time in Consumption and Residential Location in an Urban Setting," *American Economic Review,* **67** (1977), 996–1003.
[65] Hochman, O. and D. Pines, "Costs of Adjustment and Demolition Costs in Residential Construction and Their Effects on Urban Growth," *Journal of Urban Economics,* **7** (1980), 2–19.
[66] Hochman, O. and D. Pines, "Costs of Adjustment and the Spatial Pattern of a Growing Open City," *Econometrica,* **50** (1982) 1371–1391.
[67] Imai, H., "CBD Hypothesis and Economies of Agglomeration," *Journal of Economic Theory,* **28** (1982), 275–299.
[68] Isard, W., *Location and Space Economy,* Cambridge, Mass.: MIT Press, 1956.
[69] Kanemoto, Y., "Optimum, Market and Second-Best Land Use Patterns in a von Thunen City with Congestion," *Regional Science and Urban Economics,* **6** (1976), 23–32.
[70] Kanemoto, Y., *Theories of Urban Externalities,* Amsterdam: North-Holland, 1980.
[71] Karmann, A., "Spatial Barter Economies under Locational Choice," *Journal of Mathematical Economics,* **9** (1982), 259–274.
[72] Kemper, P. and R. Schmenner, "The Density Gradient for Manufacturing Industry," *Journal of Urban Economics,* **1** (1974), 410–427.
[73] Kern, C. R., "Racial Prejudice and Residential Segregation: The Yinger Model Revisited," *Journal of Urban Economics,* **10** (1981), 164–172.
[74] King, A. T., "General Equilibrium with Externalities: A Computational Method and Urban Applications," *Journal of Urban Economics,* **7** (1980), 84–101.
[75] Koopmans, T. C. and M. J. Beckmann, "Assignment Problems and the Location of Economic Activities," *Econometrica,* **25** (1957), 53–76.
[76] Lea, A. C., "Public Facility Location Models and The Theory of Impure Public Goods," *Sistemi Urbani,* **3** (1981), 345–390.
[77] LeRoy, S. F. and J. Sonstelie, "Paradise Lost and Regained: Transportation Innovation, Income, and Residential Location," *Journal of Urban Economics,* **13** (1983), 67–89.
[78] Levhari, D., Y. Oron and D. Pines, "A Note on Unequal Treatment of Equals in an Urban Setting," *Journal of Urban Economics,* **5** (1978), 278–284.
[79] Mills, D. E., "Growth, Speculation and Sprawl in a Monocentric City," *Journal of Urban Economics,* **10** (1981), 201–226.
[80] Mills, E. S., An Aggretate Model of Resource Allocation in a Metropolitan Area, *American Economic Review: Papers and Proceedings,* **57** (1967), 197–210.
[81] Mills, E. S., *Studies in the Structure of the Urban Economy,* Baltimore: Johns Hopkins, 1972.
[82] Mills, E. S., *Urban Economics,* Glenview, Illinois: Scott, Foresman Company, 1972.
[83] Mills, E. S., *Urban Economics,* second edition, Glenview, Illinois: Scott, Foresman and Company, 1980.

[84] Mills, E. S., and D. M. de Ferranti, "Market Choices and Optimum City Size," *American Economic Review, Papers and Proceedings,* **61** (1971), 340–345.
[85] Mills, E. S. and J. MacKinnon, "Notes on the New Urban Economics," *Bell Journal of Economics and Management Science,* **4** (1973), 593–601.
[86] Miron, J. R. and P. Skarke, "Non-Price Information and Price Sustainability in the Koopmans-Beckmann Problem," *Journal of Regional Science,* **21** (1981), 117–122.
[87] Mirrlees, J. A., "The Optimum Town," *The Swedish Journal of Economics,* **74** (1972), 114–135.
[88] Miyao, T., "Dynamics and Comparative Static in the Theory of Residential Location," *Journal of Economic Theory,* **11** (1975), 133–146.
[89] Miyao, T., *Dynamic Analysis of the Urban Economy,* New York: Academic Press, 1981.
[90] Mohring, H., "Land Values and the Measurement of Highway Benefits," *Journal of Political Economy,* **69** (1961), 236–249.
[91] Montesano, A., "A Restatement of Beckmann's Model of Urban Rent and Residential Density," *Journal of Economic Theory,* **4** (1972), 329–354.
[92] Moses, L. N., "Towards a Theory of Intra-Urban Wage Differentials and Their Influence on Travel Patterns," *Papers and Proceedings of the Regional Science Association,* **9** (1962), 53–63.
[93] Muth, R. F., *Cities and Housing,* Chicago: University of Chicago Press, 1969.
[94] Muth, R. F., "The Derived Demand for Urban Residential Land," *Urban Studies,* **8** (1971), 243–254.
[95] Niedercorn, J. H., "A Negative Exponential Model of Urban Land Use Densities and Its Implications for Metropolitan Development," *Journal of Regional Science,* **11** (1971), 317–326.
[96] Odland, J., "The Spatial Arrangement of Urban Activities: A Simultaneous Location Model," *Environment and Planning A,* **8** (1976), 779–791.
[97] Odland, J., "The Conditions for Multi-center Cities," *Economic Geography,* **54** (1978), 234–244.
[98] Ogawa, H., *A General Model of Urban Spatial Structure: Nonmonocentric and Multicentric,* Ph.D. Dissertation in Regional Science, University of Pennsylvania, Philadelphia, 1980.
[99] Ogawa, H. and M. Fujita, "Equilibrium Land Use Patterns in a Nonmonocentric City," *Journal of Regional Science,* **20** (1980), 455–475.
[100] O'Hara, D. J., "Location of Firms within a Square Central Business District," *Journal of Political Economy,* **85** (1977), 1189–1207.
[101] Ohls, J. C. and D. Pines, "Discontinuous Urban Development and Economic Efficiency," *Land Economics,* **51** (1975), 224–234.
[102] Oron, Y., D. Pines and E. Sheshinski, "Optimum vs. Equilibrium Land Use Pattern and Congestion Toll," *Bell Journal of Economics and Management Science,* **4** (1973), 619–636.
[103] Papageorgiou, G. J., "Spatial Externalities I: Theory," *Annals of the Association of American Geographers,* **68** (1978), 465–476.
[104] Papageorgiou, G. J., "Spatial Externalities II: Applications," *Annals of the Association of American Geographers,* **68** (1978), 477–492.
[105] Papageorgiou, G. J. and E. Casetti, "Spatial Equilibrium Residential Land Values in a Multicentre Setting," *Journal of Regional Science,* **11** (1971), 385–389.
[106] Papageorgiou, Y. Y., and J. F. Thisse, "Agglomeration as Spatial Interdependence Between Firms and Households," *Journal of Economic Theory,* 37, 1985, 19–31.

[107] Parr, J. B. and C. Jones, "City Size Distributions and Urban Density Functions: Some Interrelationships," *Journal of Regional Science,* **23** (1983), 283–307.

[108] Pines, D. "Dynamic Aspects of Land Use Pattern in a Growth City," in G. J. Papageorgiou, ed., *Mathematical Land Use Theory,* Lexington, Mass.: Lexington Books, 1976.

[109] Pines, D. and E. Werczberger, "A Linear Programming Model of the Urban Housing and Land Markets: Static and Dynamic Aspects," *Regional Science and Urban Economics,* **12** (1982), 211–233.

[110] Polinsky, A. M., "The Demand for Housing: A Study in Specification and Grouping," *Econometrica,* **45** (1977), 447–461.

[111] Radner, R., "Existence of Equilibrium of Plans, Prices, and Price Expectations in a Sequence of Markets," *Econometrica,* **40** (1972), 289–303.

[112] Richardson, H. W., *The New Urban Economics: and Alternatives,* London: Pion Limited, 1977.

[113] Richardson, H. W., "On the Possibility of Positive Rent Gradients," *Journal of Urban Economics,* **4** (1977), 60–68.

[114] Riley, J. G., "Gammaville: An Optimal Town," *Journal of Economic Theory,* **6** (1973), 471–482.

[115] Riley, J. G., "Optimal Residential Density and Road Transportation," *Journal of Urban Economics,* **1** (1974), 230–249.

[116] Robson, A. J., "Cost-Benefit Analysis and the Use of Urban Land for Transportation," *Journal of Urban Economics,* **3** (1976), 180–191.

[117] Romanos, M. C., "Household Location in a Linear Multi-Center Metropolitan Area," *Regional Science and Urban Economics,* **7** (1977), 233–250.

[118] Rose-Ackerman, S., "Racism and Urban Structure," *Journal of Urban Economics,* **2** (1975), 85–103.

[119] Rose-Ackerman, S., "The Political Economy of a Racist Housing Market," *Journal of Urban Economics,*" **4** (1977), 150–169.

[120] Sakashita, N., "Optimal Location of Public Facilities under Influence of the Land Market," Paper Presented at the European Regional Science Meeting, Budapest, forthcoming on *Journal of Regional Science* (1986)

[121] Samuelson, P. A., "Intertemporal Price Equilibrium: A Prologue to the Theory of Speculation," *Weltwirtschaftliches Archiv* 79, Hamburg: Hoffman & Campe Verlag, 1957. Reprinted in *The Collected Scientific Papers of Paul A. Samuelson,* vol. 2, MIT Press, 1966.

[122] Samuelson, P. A., "Indeterminancy of Development in a Heterogeneous Capital Model with Constant Saving Propensity," in *Essays on the Theory of Optimal Economic Growth* (Karl Shell ed.), The MIT Press. Reprinted in the *Collected Scientific Papers of Paul A. Samuelson,* vol. 3, MIT Press, 1972.

[123] Samuelson, P. A., "Thünen at Two Hundred," *Journal of Economic Literature,* **21** (1983), 1468–1488.

[124] Schnare, A. B., "Racial and Ethnic Price Differentials in an Urban Housing Market," *Urban Studies,* **13** (1976), 107–120.

[125] Schuler, R. E., "The Interaction Between Local Government and Urban Residential Location," *American Economic Review,* **64** (1974), 682–696.

[126] Schuler, R. E., "The Interaction Between Local Government and Urban Residential Location: Reply and Further Analysis," *American Economic Review,* **66** (1976), 968–975.

[127] Schulz, N. and K. Stahl, "Oligopolistic Industry Location and Local Labor Markets," Working Papers in Economic Theory and Urban Economics, No. 8306, Universitat Dortmund, Dortmund, 1983.

[128] Schweizer, U., "A Spatial Version of the Nonsubstitution Theorem," *Journal of Economic Theory,* **19** (1978), 307–320.
[129] Schweizer, U. and P. Varaiya, "The Spatial Structure of Production with a Leontief Technology," *Regional Science and Urban Economics,* **6** (1976), 231–251.
[130] Schweizer, U. and P. Varaiya, "The Spatial Structure of Production with a Leontief Technology—II: Substitute Techniques," *Regional Science and Urban Economics,* **7** (1977), 293–320.
[131] Schweizer, U., P. Varaiya, and J. Hartwick, "General Equilibrium and Location Theory," *Journal of Urban Economics,* **3** (1976), 285–303.
[132] Scotchmer, S., "Hedonic Prices, Crowding and Optimal Dispersion of Population," Paper presented at the Table Ronde Models Economiques De La Localization et des Transports, Paris, 1982.
[133] Smith, T. R., and G. J. Papageorgiou, "Spatial Externalities and the Stability of Interacting Populations Near the Center of a Large Area," *Journal of Regional Science,* **22** (1982), 1–18.
[134] Solow, R. M., "Congestion, Density and the Use of Land in Transportation," *The Swedish Journal of Economics,* **74** (1972), 161–173.
[135] Solow, R. M., "Congestion Cost and the Use of Land for Streets," *Bell Journal of Economics and Management Science,* **4** (1973), 602–618.
[136] Solow, R. M., "On Equilibrium Models of Urban Locations," in J. M. Parkin, ed., *Essays in Modern Economics,* London: Longman, 1973.
[137] Solow, R. M., and W. S. Vickrey, "Land Use in a Long Narrow City," *Journal of Economic Theory,* **3** (1971), 430–447.
[138] Starrett, D., "Market Allocation of Location Choice in a Model with Free Mobility," *Journal of Economic Theory,* **17** (1978), 21–37.
[139] Strotz, R. H., "Urban Transportation Parables," in J. Margolis, ed., *The Public Economy of Urban Communities,* Baltimore: Johns Hopkins, 1965.
[140] Sullivan, A. M., "A General Equilibrium Model with External Scale Economies in Production," *Journal of Urban Economics,* **13** (1983), 235–255.
[141] Tauchen, H., "The Possibility of Positive Rent Gradients Reconsidered," *Journal of Urban Economics,* **9** (1981), 165–172.
[142] Tauchen, H. and A. D. Witte, "An Equilibrium Model of Office Location and Contact Patterns," *Environment and Planning A,* **15** (1983), 1311–1326.
[143] Tauchen, H. and A. D. Witte, "Socially Optimal and Equilibrium Distributions of Office Activity: Models with Exogenous and Endogenous Contacts," *Journal of Urban Economics,* **15** (1984), 66–86.
[144] ten Raa, T., "The Distribution Approach to Spatial Economies," *Journal of Regional Science,* **24** (1984), 105–117.
[145] Tiebout, C. M., 1956, "A Pure Theory of Local Expenditures," *Journal of Political Economy,* **64** (1956), 416–424.
[146] Thisse, J. F. and H. G. Zoller, *Locational Analysis of Public Facilities,* Amsterdam: North-Holland, 1983.
[147] Von Thünen, J. H., *Der Isolierte Staat in Beziehung auf Landwirtschaft und Nationalekonomie,* Hamburg, 1826.
[148] Wheaton, W. C., "A Comparative Static Analysis of Urban Spatial Structure," *Journal of Economic Theory,* **9** (1974), 223–237.
[149] Wheaton, W. C., "On the Optimal Distribution of Income Among Cities," *Journal of Urban Economics,* **3** (1976), 31–44.
[150] Wheaton, W. C., "Income and Urban Residence: An Analysis of Consumer Demand for Location," *American Economic Review,* **67** (1977), 620–631.

[151] Wheaton, W. C., "Urban Residential Growth under Perfect Foresight," *Journal of Urban Economics,* **12** (1982), 1–21.

[152] Wheaton, W. C., "Urban Spatial Development with Durable but Replacable Capital," *Journal of Urban Economics,* **12** (1982), 53–67.

[153] White, M. J., "Firm Suburbanization and Urban Subcenters," *Journal of Urban Economics,* **3** (1976), 323–343.

[154] Wingo, L. Jr., *Transportation and Urban Land,* Washington, D.C.: Resources for the Future, 1961.

[155] Yamada, H., "On the Theory of Residential Location: Accessibility, Space, Leisure and Environmental Quality," *Papers of the Regional Science Association,* **29** (1972), 125–135.

[156] Yang, C. H. *Urban Spatial Structure with Local Public Goods: Optimum and Equilibrium,* Ph.D. dissertation, University of Pennsylvania, Philadelphia, 1980.

[157] Yang, C. H., and M. Fujita, "Urban Spatial Structure with Open Space," *Environment and Planning A,* **15** (1983), 67–84.

[158] Yellin, J., "Urban Population Distribution, Family Income, and Social Prejudice," *Journal of Urban Economics,* **1** (1974), 21–47.

[159] Yinger, J., "Racial Prejudice and Racial Residential Segregation in an Urban Model," *Journal of Urban Economics,* **3** (1976), 383–396.

General Equilibrium in Space and Agglomeration

Urs SCHWEIZER†

University of Bonn, Bonn, FRG

0. INTRODUCTION

This survey reviews some of the literature on space in general equilibrium theory. The material to be covered has been selected according to the following criteria. First, general equilibrium theory is meant to deal not only with the existence or nonexistence of Walrasian equilibrium in spatial economies. The number of contributions on this topic would not be sufficient to make for a full survey. The subject is rather extended to include first and second welfare theorems for models of location choice. The viewpoint, however, will be that of Paretian welfare economics where allocations are defined to be (Pareto) efficient if utility of all consumers cannot be uniformly increased. Mirrlees [38] has pointed to the fact that maximizing a social welfare function might well require that equals have to be treated unequally (see, e.g., [12], [36], [20] and [65] for further elaboration of that theme). In the present survey, however, allocations will only be evaluated according to the Pareto criterion, mainly because I believe social welfare functions to rest on shaky ground as far as their foundation is concerned. If their use in spatial economies leads to disturbing conclusions such as equals must be treated unequally, this topic should be covered rather by a survey on welfare functions than on spatial equilibrium. Second, except for the requirement that the sum of marginal rates of

† The author wants to thank Richard Arnott, Masahisa Fujita, John Hartwick and David Wildasin for helpful comments on an earlier draft.

substitution over all consumers should equal the marginal rate of transformation, traditional equilibrium theory has little to offer towards understanding the supply mechanisms for public goods. It is within the spatial context that one can hope for some more interesting results. Therefore, part of the literature on local public expenditures dealing with consumer mobility and land use will be reviewed as well. In particular, the notion, shape, existence and nonexistence of Tiebout equilibrium will be discussed for the case where limits to agglomeration economies arise from land constraints. We do not consider other congestion externalities because they are treated elsewhere (see [27] and [28]), nor do we discuss so-called politico-economic equilibria where the level of local public spending is determined by some voting mechanism. The reader interested in the latter topic should read [64]. Third, three major causes of urban agglomeration have been identified in the literature—diversity of the resource base, economies of scale in the provision of public services, and agglomerative economies of scale in private production (see [39] or [2] among others). While this survey deals with the first two causes, economies of scale in private production will not be discussed, mainly because general equilibrium theory does not really seem ready to plunge into that topic within an aspatial let alone a spatial framework. The reader is referred to Serck-Hanssen [50], Starrett [53] and Vickrey [62] for contributions to this subject. To summarize, this survey aims at presenting the theory of Pareto efficient allocations in space and of equilibrium in models with location choice and public investment.

Except for the locationally separated markets without location choice literature covered in Section 1, the relevant literature for this survey will be reviewed along the lines of a unifying framework as presented in Section 2.1 below. The main ideas appearing in the existing literature for quite different models will be adopted to our single framework. Using this approach, the various findings are more conveniently compared and potential gaps become apparent indicating where further research is needed most urgently. Our model is by no means as general as current knowledge would allow. We have rather searched for the simplest framework yet one still rich enough to discuss all important aspects of the theory.

The survey is organized as follows. In Section 1, the classical partial equilibrium analysis, which predicts the intersection of the demand and supply curves to be the equilibrium allocation, is extended to the case of locationally separated markets for qualitatively the same commodity. At every market location a local demand and supply curve resulting from economic behavior of agents living in those locations is specified. The shipment of commodities between markets is costly. In equilibrium, all markets have to clear at prices which prevent arbitrage within the transport sector. This is the so-called Enke-Samuelson problem of locationally separated markets (see [13] and [41]) which, however, has been studied before by Cournot (see [17]). Samuelson [41] has proposed to find a solution by looking at an associated maximum problem where equilibrium prices are shown to emerge as dual variables. Takayama and Judge [56, 57] have devised computational algorithms for solving the associated maximum problem which turn out to be very useful for empirical partial equilibrium analysis.

In Section 2 the basic model, which will be discussed for the remainder of the survey, is introduced. Following Debreu [10], space is assumed to be divided into a finite set of elementary regions which are arbitrarily numbered according to location. These elementary regions are chosen small enough such that points of them need not be further distinguished. Put differently, each location (out of a finite set) can be seen as being endowed with some fixed amount of land (in our model) for residential use by consumers who could be of different types. The commodity space is enlarged in the familiar way (see [10]) by spatially indexed commodities. To deal with location choice properly, care must be taken in specifying preference orderings over the enlarged commodity space. It is here where we have to depart substantially from the Arrow–Debreu model. Consumers' choice must be viewed as a two-stage process. At the first stage, consumers will decide where to reside. At the second stage, for every possible location choice, consumers must be able to evaluate consumption alternatives. In other words, a different preference ordering over the enlarged commodity space must be specified for each choice of location. In an ordinary Arrow–Debreu model, there exists just one preference ordering for each consumer, whereas in a model of location choice there have to be as many orderings as there are different residential locations. This approach captures the idea that utility derived from consuming land at any particular location crucially depends on where the consumer lives. In extreme, land at any other than the residential location might be of no value to that consumer. This extreme case is the assumption commonly imposed by the literature. But less restrictive assumptions could probably be explored as well. To summarize, enlarging the commodity space by spatially indexed commodities is an ingenious device to deal with land and transportation in an Arrow–Debreu world. But this is not enough to describe location choice endogenously. It is consumption as a two-stage process which has to be added and which makes existence of equilibrium with location choice a difficult subject.

Our basic model allows further for governmental authorities providing the economy with public investments in the form of both local and national public goods. Public expenditures will have to be covered from tax revenues. In particular, we shall examine what kind of taxation would be needed to sustain Pareto efficient solutions. Costly transportation of non-land commodities, as discussed in Section 1, will neither be further examined in Section 2 nor in later sections. It is rather assumed that there exists only one such commodity and that it can be shipped between locations at no costs. Except for burdensome notation, adding costly transportation of the non-land commodity to our model should not lead to further difficulties. Finally, the model has no production of private commodities. Again, including fully divisible production as is done in traditional general equilibrium theory would cause few problems. In the presence of indivisibilities, however, the price mechanism is quite likely to break down as has been shown by Koopmans and Beckmann [33]. Their findings are reported in Section 3 not within their original framework but rather by adopting them to our model of residential choice.

The topics to be dealt with in this survey can be classified according to different notions of efficiency. First, for what we call the *assignment problem,* (Pareto) efficiency will be explored with a fixed supply of public goods. We distinguish between the *integer* and the *fractional* assignment problem according to whether the underlying assignment of people to locations has to be integer or not. The fractional case, loosely speaking, corresponds to a large number of consumers. Here, the non-convexity arising from viewing consumption choice as a two-stage process can be shown to be smoothened out and, hence, not to prevent existence of equilibrium. Section 2.2 gives a set of conditions in terms of prices and payments which are sufficient for an allocation to solve the fractional assignment problem. This result will be referred to as the *first welfare theorem* (for the assignment problem). It follows that any allocation solving the integer but not the fractional assignment problem cannot be sustained in the sense of this theorem. Along these lines, nonexistence of spatial equilibrium in the presence of indivisibilities will be discussed in Section 3, leading to the conclusion that, in general, equilibrium can be hoped for to exist only in the case of fractional assignments. In Section 4.1, it is shown that indeed any solution of the fractional assignment problem must necessarily be sustainable in the sense of the first welfare theorem. For obvious reasons, this result is called the *second welfare theorem* which holds for the fractional assignment problem. In Section 4.2, the assignment problem is considered from the view of partial equilibrium analysis. As in the case of locationally separated markets discussed in Section 1, partial equilibrium allocations can be shown to exist by solving an associated maximum problem. The approach, though successfully applied to systems of separated markets, has hardly been explored for models of location choice. More general results on existence of (fractional) equilibrium based on fixed-point theorems will not be explicitly dealt with in this survey.

Sandler and Tschirhart [43] in their survey on the economic theory of clubs distinguish between a "within club" model which will be dealt with in Section 4.1 and 4.2 and a "total economy" model which will be discussed in Section 4.3. The distinction proves useful in clearing up some confusion which has arisen in the literature (see Helpman and Hillman [25]). For example, the original concept of clubs as pioneered by Buchanan [9] corresponds to the "within club" view where a representative member's utility is considered when optimal decisions are to be derived for the club. Welfare effects on nonmembers are neglected. The "total economy" model, on the other hand, takes the given national population fully into account. For this approach, our model of location choice must be seen as part of a larger setup such as a single agglomeration relative to the whole country. From the agglomeration's viewpoint, the optimal size and structure of population is reached if the utility per head cannot be uniformly increased. Following Schweizer [46], such an allocation will be called *club efficient*.

Club efficiency must be carefully distinguished from efficiency at the "total economy" level for basically three reasons. First, even at an unlimited supply of land, it may not be possible to subdivide the given national population into clubs all of which are of optimum size. Here, an integer problem arises with respect to the number of agglomerations beyond what can be repaired by resorting to fractional assignments (unless we are prepared to deal with a continuum instead of a finite number of clubs!). Second, land endowments may support a limited number of agglomerations only. If the national population exceeds what could be absorbed by these agglomerations at optimum size, it would certainly not be desirable to insist on club efficient allocations leaving some people to live nowhere. Third, the utility per head obtained in agglomerations of optimum size depends on quality and endowment of the resource base. Diversity of the base leads to different levels of utility for consumers of the same type and, thus, would be inconsistent with equilibrium under free mobility.

In spite of these problems, the club approach seems interesting enough to merit a full discussion in the present survey, in particular because it allows for a complete set of welfare theorems. In Section 4.3, a first and second *club welfare theorem* (for the club assignment problem) will be established which leads to a general version of the so-called *Henry George Theorem* on the relation between aggregate land rent and public expenditures (see Arnott [1], Arnott and Stiglitz [3] and Schweizer [45, 46] among others). The two club welfare theorems allow us to characterize the set of utility allocations at which solutions of the club assignment problem exist. In particular, it is shown, that some solutions can only be realized by mixed clubs whereas, in other cases, a single consumer type is able to sustain a solution on its own.

In Section 5, the assumption of a fixed supply of public goods will be dropped. In Section 5.1, the efficient supply is first studied for a given assignment of consumers to locations. The situation corresponds to what we call the *Lindahl-Samuelson problem.* A first and second welfare theorem holds for this problem as well. Since the assignment is kept fixed, these theorems are essentially aspatial. For the so-called *first best problem* optimum supply of public goods and assignment of consumers have to be determined simultaneously. While this problem still allows for a second welfare theorem (necessary condition for optimality), a corresponding first welfare theorem, in general, cannot be established. Both a variable level of public spending and assignment may give rise to nonconvexities which allow nonoptimal allocations to become sustainable. Section 5.2 deals with the club version of the first best problem which is shown to be closely related to Bewley's [8] notion of Tiebout equilibrium. In this section, we use a two-step approach to the first best problem by first dealing with the assignment problem at a fixed supply of public goods. This proves very useful for explaining why Tiebout equilibria will quite likely fail to exist and why, if they exist, consumers might well choose to live in mixed communities.

In summary, our understanding of the assignment problem turns out to be far more complete than that of the first best problem. Beyond classification of current knowledge, however, the two-step approach may also reflect positive aspects of fiscal decentralization. It seems plausible from casual observation that level and structure of public spending are more stable over time than residential patterns due to the way in which changes occur. For public spending, sluggish political mechanisms may lead to slow changes only whereas, for residential patterns, changes occur rather quickly as a consequence of consumers' decisions. The assignment problem then seems to correspond, in the sense of positive theory, to the adjustment of residential choice to a given pattern of public spending. The question of which factors determine such patterns has to be the subject of an as of yet largely unexplored positive theory of decision taking by decentralized governmental authorities (see [64]). In addition, the negative findings as far as the existence of Tiebout equilibria are concerned suggest that some federal coordination is needed so that competition among local governments will work properly. Equilibrium, in general, can only be defined with respect to a given scheme of coordination. While Richter [40] and Greenberg [15] offer results in this direction, further research would be most desirable in this area.

1. SPATIALLY SEPARATED MARKETS

In a rather famous paper, Samuelson [41] takes up the subject, studied before by Cournot and later by Enke, of how to incorporate distance and costly transportation into (partial) equilibrium analysis. In this model, markets for a given commodity are assumed to exist at spatially separated locations $k = 1, 2, \ldots, K$. It costs c_{kl} units of money (or of a composite good) to ship one unit of the commodity from location k to l. At every location, a demand and supply curve is specified. Demand at k is denoted by $p_k = D_k(x_k)$ whereas, to keep in the spirit of later sections, supply at k is assumed to be exogenously given at level X_k^0. A *partial equilibrium* consists of a commodity price p_k and a level x_k^* of demand for all locations k and of a nonnegative commodity flow s_{kl}^* from k to l such that

$$x_k = X_k^0 + \sum_l (s_{lk}^* - s_{kl}^*) \tag{1.1.}$$

$$p_l \leq p_k + c_{kl} \tag{1.2.}$$

and, for all k, $p_k = D_k(x_k^*)$. Equality has to prevail in (1.2.) if the flow s_{kl}^* from k to l is positive. Condition (1.1.) requires that local demand equals supply, whereas (1.2.) prevents arbitrage within the transport sector. If commodities have to be shipped the difference in prices must cover transport costs. These are the obvious requirements for equilibrium.

In the aspatial model of only one market, the equilibrium allocation is simply found as the intersection of demand and supply curve. For the spatial setup, however, the existence of equilibrium can only be established by more elaborate methods. The method proposed by Samuelson involves maximizing the net social payoff defined as the sum of consumers' plus producers' surplus minus resources spent on transportation (associated maximum problem). Equilibrium prices then appear as variables dual to the balancing constraints of the different locations. Details of the approach can be worked out in the following way.

Demand D_k at k is given the meaning of marginal utility from which utility $V_k(x_k)$ is obtained by solving the differential equation $V_k'(x_k) = D_k(x_k)$ with initial value $V_k(0) = 0$. For any allocation $x = (\ldots, x_k, \ldots)$ and $s = (\ldots, s_{kl}, \ldots)$, the net social payoff amounts to

$$\sum_k V_k(x_k) - \sum_k \sum_l c_{kl} s_{kl}.$$

The associated maximum problem consists of maximizing this

payoff over all allocations satisfying the balancing constraints

$$x_k \leq X_k^0 + \sum_l (s_{lk} - s_{kl}). \tag{1.3.}$$

The solution (x^*, s^*) of the associated maximum problem which generally exists can easily be shown to be sustained by the shadow prices p_k dual to (1.3.) as a partial equilibrium. Therefore, equilibrium must exist and, as a matter of fact, can be computed for cases far more general than ours. Takayama and Judge [56, 57] have introduced algorithms for the multiproduct case with linear demand and supply schedules which have been successfully applied for empirical studies (see [32] among many others). The method has since been substantially refined (see [52, 58, 59, 60, 17, 18]). The price to be paid, however, will always be partial instead of general equilibrium analysis. More precisely, to justify the approach from the general equilibrium viewpoint some rather restrictive assumptions on preferences have to be imposed. Utility $U_k(x_k, y_k)$ defined on the consumption of both x_k the commodity under consideration and the composite good y_k must be separable in the sense of $U_k(x_k, y_k) = V_k(x_k) + y_k$. In this case, consumers' surplus is a valid utility indicator and the given partial equilibrium (x^*, s^*, p) can be extended to a general equilibrium allocation in the following way.

Let Y_k^0 denote the endowment with the composite good of consumers living at location k. The corresponding equilibrium level y_k^* of consumption must be calculated from the budget equation

$$p_k x_k^* + y_k^* = p_k X_k^0 + Y_k^0. \tag{1.4.}$$

Provided that y_k^* is nonnegative and the preferences allow for representation by ordinal utility functions separable in the above sense, the consumption bundle (x_k^*, y_k^*) maximizes utility over the proper budget set because, by definition of a partial equilibrium, the price p_k at k equals marginal utility $V'_k\ (x_k^*)$. Since utility is maximized, the extended allocation (x^*, y^*, s^*, p) will indeed be a general equilibrium. For nonseparable preferences, however, the existence of equilibrium can no longer by established by solving an associated maximum problem. Here, fixed point theorems are needed as has been shown by Karmann [29, part II] for a rather general model which fully covers the simple case discussed in the present section.

2. INTRODUCING LOCATION CHOICE

In this section, a model focussing on aspects of location choice is introduced which serves as the basic framework for the remaining part of this survey. After presenting the model, Section 2.1 recalls in detail classification of problems as outlined in the introduction. Section 2.2 contains a first welfare theorem for economies with location choice as well as a discussion of its notion of sustainability.

2.1. The model: classification of problems

In the basic model, every location $k = 1, 2, \ldots, K$ is endowed with a fixed amount X_k^0 of a commodity which, in contrast to Section 1, is no longer assumed to be mobile and which we refer to as land. Since many papers to be reviewed deal with devoting resources to public investments, a supply $z = (\ldots, z_k, \ldots)$ of public goods has to be introduced where z_k denotes public investments at location k. Total costs (public expenditures) amount to $C = C(z)$ units of the composite good. The set of consumers is divided into classes of different types. Consumer types are denoted by $i = 1, 2, \ldots, I$ whereas N_i^0 is the number of consumers of type i at the national level (total economy). An *assignment* $n = (\ldots, n_{ik}, \ldots)$ of consumers to locations describes the number n_{ik} of consumers of type i living at location k. For the "total economy" approach (see introduction), an assignment must satisfy the population constraint

$$\sum_k n_{ik} = N_i^0 \tag{2.1.}$$

for all types, whereas, for the "within club" model, the constraint is not taken into account. Except for the club viewpoint, all land is assumed to be owned by consumers. Each consumer of type i owns x_{ik}^0 units of land at location k, hence

$$X_k^0 = \sum_i N_i^0 x_{ik}^0 \quad \text{for all } k.$$

Production of private commodities will not be considered explicitly. Instead, we assume that a consumer of type i living at k is endowed with y_{ik}^0 units of the composite good. These endowments are listed net of commuting costs and, for that reason, appear with the index k of residential choice. By assumption, the composite good can be shipped between locations at zero cost. Therefore, the composite good need not be indexed spatially. The (enlarged) commodity space is R^{K+1} as far as private commodities are concerned. In a traditional Arrow–Debreu framework, preferences of every consumer would simply be defined over the enlarged commodity space. If such preferences were convex as commonly assumed, every consumer would tend to consume land at every location. This causes severe difficulties for interpretation. In particular, the idea of endogenous location choice would certainly not be captured by this approach. Consumer choice must rather be seen as a two-stage process (see introduction) which means that, for any choice of location, a different preference ordering has to be specified. It then becomes clear which of the land is used for residential purposes and which, if any, is devoted to other activities. In the following, however, it is assumed for simplicity that land at other than the residential location does not create utility for that consumer. In other words, for any choice of location, the corresponding utility function depends only on land consumption at that location and on consumption of the composite good but not on land consumption elsewhere. It is this two-stage approach to consumer choice which leads to nonconvexities and, therefore, makes existence of equilibrium a more difficult issue than in the traditional Arrow–Debreu model. For every consumer of type i residing at location k, private consumption of land and the composite good is denoted by x_{ik} and y_{ik}, respectively, whereas his utility function is denoted by $U_{ik}(x_{ik}, y_{ik}, z)$. Since location choice appears as an index of the utility function, the model is general enough to allow for consumers having intrinsic preferences for particular locations.

Due to free mobility, consumers of the same type have to obtain the same utility level in equilibrium. In this sense, *utility allocations* $u = (\ldots, u_i, \ldots)$ prescribe a unique utility level u_i for every consumer type i. An *allocation* for the given utility allocation u consists of: an assignment $n = (\ldots, n_{ik}, \ldots)$, vectors $x = (\ldots, x_{ik}, \ldots)$ and $y = (\ldots, y_{ik}, \ldots)$ of consumption of land and the composite good, a supply $z=(\ldots, z_k, \ldots)$ of public investments and, a level S of exports such that (2.1.) as well as

$$U_{ik}(x_{ik}, y_{ik}, z) \geq u_i \quad \text{for all } i \text{ and } k \tag{2.2.}$$

$$X_k^0 \geq \sum_i n_{ik} x_{ik} \quad \text{for all } k, \text{ and} \tag{2.3.}$$

$$S \leq \sum_i \sum_k n_{ik}(y_{ik}^0 - y_{ik}) - C(z) \tag{2.4.}$$

hold. (2.2.) requires that consumers obtain the utility level prescribed by the given utility allocation. (2.3.) is the land constraint for location k. Since the composite commodity can be shipped at zero cost, the right-hand side of (2.4.) is excess supply of the composite commodity. In the following, to allow for extra generality, excess supply will not be required to vanish. If it is positive, excess supply or what we propose to call the level S of export corresponds to the consumers' aggregate rent-paying ability which might leave the system to be absorbed by absentee landlords or land developers. The case of zero exports $S=0$ corresponds to an allocation for the fully closed economy whereas a negative value means that the proper number of units of the composite good have to be supplied to the economy from the outside.

Various notions of efficiency will be discussed in subsequent sections. They all have in common that the level S of exports must attain its maximum over a set of allocations consistent with the prescribed utility allocation u. The approach deviates slightly from convention where, instead, utility allocations would qualify as efficient if they cannot be improved upon at a fixed level of exports. For preferences monotonic in the composite good, however, our notion is if anything more general and, more important, notationally simpler to handle. The normative view just described is obviously related to that taken by Herbert and Stevens [26] (see also [47]). Having this in mind, the different notions of efficiency are classified according to the set of allocations over which to maximize the level of exports S. For the *assignment problem,* the supply of public goods z is kept fixed at some possibly suboptimal value. An *integer* and a *fractional* assignment problem will be distinguished depending on whether the underlying assignment of people must be integer or not. The "within club" view of the above situation will be referred to as the *club assignment problem.* Here, the level of exports has to be maximized at a given supply of public goods while the population constraint (2.1.) need not be observed. For the *Lindahl-Samuelson* problem, the level of exports is maximized at a fixed assignment n with variable supply z and, for the *first best problem,* both n and z are to be determined endogenously. Alternatively, let us define $s(u, n, z)$ as the maximum level of exports over all allocations with underlying assignment n and public spending z yielding the given utility allocation u. The above classification then rests on whether the maximum of the function $s(u, n, z)$ is taken with respect to n, z or both and whether the population constraint is binding or not.

2.2. The first welfare theorem

In this section, the first welfare theorem of general equilibrium theory is extended to our basic model of location choice. Any allocation sustainable by land rents and payments in the sense of the following theorem is shown to solve the fractional assignment problem. Since the level z of public investments will be kept fixed throughout this section as well as Sections 3 and 4, we will omit writing z as an explicit parameter.

THEOREM 1 *For a given utility allocation u, the allocation (n^A, x^A, y^A, S^A) solves the fractional assignment problem if there exists vectors $r = (r_1, \ldots, r_K) \geq 0$ of land rents and $R = (R_1, \ldots, R_I)$ of payments such that*

$$S^A + \sum_i N_i^0 R_i + C = \sum_k r_k X_k^0 \tag{2.5.}$$

and, for any consumption bundle with $U_{ik}(x_{ik}, y_{ik}) \geq u_i$,

$$r_k x_{ik} + y_{ik} \geq y_{ik}^0 + R_i. \tag{2.6.}$$

Proof Take any allocation (n, x, y, S) with a possibly fractional assignment n. It then follows from (2.1.), (2.3.), (2.4.) and (2.6.)

that

$$S+\sum_i\sum_k n_{ik}y_{ik}+C+\sum_i N_i^0R_i\le\sum_i\sum_k n_{ik}(y_{ik}^0+R_i)$$
$$\le\sum_i\sum_k n_{ik}(r_kx_{ik}+y_{ik})$$
$$\le\sum_k r_kX_k^0+\sum_i\sum_k n_{ik}y_{ik}$$

hence, by using (2.5.), $S\le S^A$. *Q.E.D.*

Before discussing the meaning of this theorem, we state the following corollary. Its proof, similar to that of the theorem, is left as an exercise to the reader.

COROLLARY 1.1 *For any location k, it follows that* $r_kX_k^0=r_k\sum_i n_{ik}^Ax_{ik}^A$. *Moreover, for any consumer type i present at* $k(n_{ik}^A>0)$, *the budget equation* $r_kx_{ik}^A+y_{ik}^A=y_{ik}^0+R_i$ *holds.*

Sustainability in the sense of Theorem 1 allows for the following interpretation. Every consumer of type i residing at location k has a budget $y_{ik}^0+R_i$ in terms of the composite good to spend on consumption. The first component of income corresponds to initial endowments net of commuting costs whereas the second component R_i is best seen as lump-sum payment received (to be paid if negative) irrespective of residential choice by consumers of type i. It follows from (2.6.) and the above corollary that consumers maximize utility (except for zero incomes where expenditures are minimized). Moreover, (2.5.) shows that aggregate land rents cover public expenditures C, exports S^A and the sum $\sum N_i^0R_i$ needed to sustain the above scheme of payments. Any Walrasian equilibrium for a closed economy ($S^A=0$) with no public spending ($C=0$) is, of course, sustainable in the sense of Theorem 1 where the payment R_i received by consumers of type i just amounts to the value $\sum_k r_kx_{ik}^0$ of their land endowments. The theorem then shows that Walrasian equilibria, in particular, solve the fractional assignment problem for the proper utility allocation. This is the first welfare theorem expressed for our model of location choice.

Sustainability in the sense of Theorem 1 can also be seen as resulting from *utility taking* behaviour of local land developers (see [55]) or of immobile land owners (see e.g. [45] or [63]). By definition, such local agents perceive the utility allocation as given by what consumers would gain from moving out of their local domain. They then attempt to maximize profits (or residual rents) by forming a community of proper size. The following corollary shows that, no matter what instruments they have at their disposal, profits cannot be increased beyond what they obtain under the sustainable allocation of Theorem 1.

COROLLARY 1.2 *For every location k, the sum* $\sum_i n_{ik}(y^0_{ik} + R_i - y_{ik})$ *attains its maximum subject to* (2.2.) *and* (2.3.) *at* $(n_k, x_k, y_k) = (n^A_k, x^A_k, y^A_k)$.

Proof It follows from (2.2.), (2.3.) and (2.6.) that $\sum_i n_{ik}(y^0_{ik} + R_i - y_{ik}) \leq r_k X^0_k$, whereas Corollary 1.1 implies that

$$\sum_i n^A_{ik}(y^0_{ik} + R_i - y^A_{ik}) = r_k X^0_k. \quad Q.E.D.$$

As a final comment on Theorem 1, we point out that any allocation sustainable at zero payments ($R = 0$) must necessarily solve the fractional club assignment problem. The proof is similar to that of Theorem 1 and need not be reproduced here. This result, more fully discussed in Section 4.3, will be referred to as the first club welfare theorem.

3. INTEGER ASSIGNMENT

In a seminal paper [33], Koopmans and Beckmann investigate the problem of assigning indivisible plants to an equal number of locations. Every location can be occupied by only one plant. Revenues generated by plants possibly depend on the location to which they are assigned. For the *linear assignment problem,* costs of transportation between plants are ignored. Total revenues are shown to attain their maximum at an (integer) assignment sustainable by a system of land rents as a market equilibrium. As a consequence, any solution of the integer assignment problem must necessarily solve its fractional version as well. For the more general *quadratic assignment problem,* however, where interplant transportation is taken into account, this need no longer be the case. Koopmans and Beckmann present examples where none of the integer assignments can be sustained as equilibrium and, hence, where equilibrium does not exist. The examples are chosen such that no solution of the integer quadratic assignment problem solves the fractional version and, for that reason, integer solutions will not be sustainable.

These rather negative findings concerning the functioning of competitive mechanisms in the presence of distance and indivisibililties have disturbed quite a few authors. Heffley [21] offers a necessary condition for the existence of equilibrium which rules out the class of examples considered by Koopmans and Beckmann. Unfortunately, his condition is not sufficient for sustainability of integer solutions, leaving the question of existence unanswered. [19], [22] and [37] attempt to moderate the negative conclusion by Koopmans and Beckmann along similar lines. Despite these efforts, nonexistence of equilibrium remains a serious problem. This view has been reinforced by Starrett [54] who introduces a method of quantifying the disequilibrium incentive associated with any integer allocation and by Hamilton [16] who ascribes the inability of the land market to work properly to the failure of the market to correctly price accessibility.

The theory of cooperative games has the notion of core allocations as a solution concept. It is well-known from general equilibrium theory (see Debreu and Scarf [11]) that every competitive equilibrium leads to a core allocation but that, on the other hand, core allocations might exist which cannot be sustained as an equilibrium. In other words, the core need not be empty, even if no equilibrium exists. Such a possibility has led some authors to look at the assignment problem from the view of game theory. Shapley and Shubik [51] deal with the linear assignment problem where the problem of nonexistence does not arise and where the sets of equilibria and the core allocations turn out to coincide. Heffley [23] attempts to approach the quadratic assignment problem in a similar way. His notion of core, again, leads to a full equivalence between core and equilibrium allocations and, hence, does not resolve the issue of nonexistence. His definition, however, seems rather unconventional because prices enter the specification of the game's characteristic function. Schotter [44] proceeds more in line with traditional game theory by assigning to any coalition a value based on the worst of all possible states of the world. The approach obviously leads to a bias in favor of a nonempty core by making blocking harder. Core allocations can then be hoped for to exist even if competitive equilibria do not.

Nonexistence of equilibrium will now be discussed within our basic model of location choice. While Beckmann [5] has offered an interpretation of the quadratic assignment problem in terms of residential choice, the example proposed below makes clear that nonexistence is not only due to the interaction between consumers to be assigned but also to the mere assumption of indivisibility. For the Starrett [54] approach to work, however, interaction between agents turns out to be crucial as the following example clearly indicates.

EXAMPLE Suppose there are just two locations ($k=1,2$) and two consumers both of the same type. For simplicity, utility is assumed to be separable and independent of residential choice and, hence, can be written as $U_k(x_k, y_k) = V(x_k) + y_k$. Initial (gros) endowment of the composite good of each consumer is denoted by y^0. Two specifications of the above model will be considered. The first leads to a situation where equilibrium fails to exist because of indivisibilities. Here, existence of equilibrium can be restored if the number of consumers is large enough (continuum of consumers). For the second specification, where there is interaction between the two consumers, non-existence seems to be more deeply rooted. Following Starrett [54], it is shown that the average incentive to move to the other location is positive and equal in size to commuting costs. At equilibrium, such incentives should be negative and hence equilibrium again will not exist.

First specification. Here, the example will be interpreted as a discrete version of a monocentric city with one inner ($k=1$) and one outer ($k=2$) location. Land at the inner location is more scarce, i.e. $X_1^0 < X_2^0$. Commuting costs from the outer to the inner location amount to t units of the composite good which means that $y_1^0 = y^0$ and $y_2^0 = y^0 - t$. The parameter configuration is chosen such that the assignment $n^I = (1, 1)$ where every consumer occupying exactly one of the two locations is the best integer assignment. This is easily shown to be the case for values of t for which

$$V(X_1^0) + V(X_2^0) - 2V(X_1^0/2) \geq t \geq 2V(X_2^0/2) - V(X_1^0) - V(X_2^0).$$

All land is devoted to residential use and therefore, at any solution of the integer assignment problem, land consumption of the consumer living at the inner and the outer location must be, respectively, $x_1^I = X_1^0$ and $x_2^I = X_2^0$, whereas consumption of the numeraire good is to be calculated from

$$V(X_1^0) + y_1^I = V(X_2^0) + y_2^I = u$$

where u is the prescribed utility level. Suppose that no resources are spent on public investments. The level S^I of exports at the integer solution then amounts to

$$S^I = 2y_1^0 - t - y_1^I - y_2^I = 2y_1^0 - t - 2u + V(X_1^0) + V(X_2^0).$$

It is not difficult to see that the integer solution does not solve the fractional assignment problem except for the unlikely case that the consumer's surplus from moving out (area ABCD in Figure 1) is just equal to commuting costs t (see Section 4.2). It then follows from Theorem 1 that, generically, the best integer solution cannot

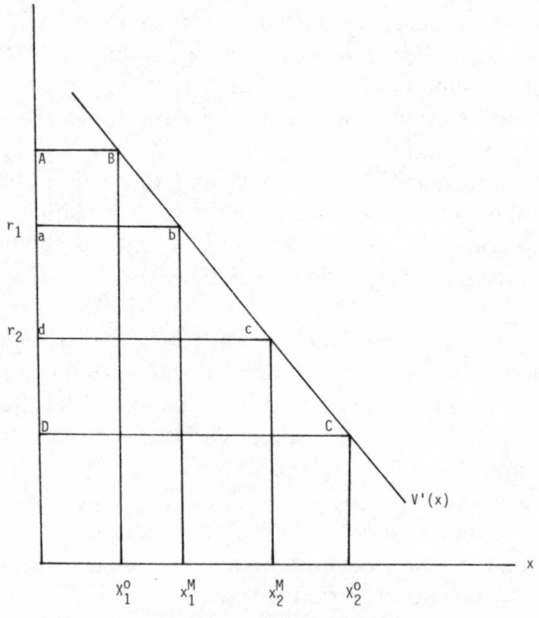

FIGURE 1 Nonexistence of equilibrium.

be sustained in the theorem's sense and, hence, equilibrium does not exist.

Second specification. For the second specification, the spatial setup is assumed to be fully symmetric. Land endowments are the same at both locations, i.e. $X_1^0 = X_2^0$. Each consumer has to see the other on a regular schedule. Commuting costs amounting to t units of the composite commodity arise for each consumer if the two live at different locations (as they will by assumption). Consumption of the consumer living at location k is denoted by (x_k, y_k). Starrett's incentive I_1 for the consumer at location 1 to move to the other location is

$$I_1 = r_1 x_1 + y_1 - (r_2 x_1 + y_1 - t).$$

If I_1 is positive, the consumer could afford the same consumption bundle (x_1, y_1) at less expenditures. Therefore, in equilibrium, he would not remain at location 1. Similarly, the incentive I_2 of the consumer living at $k = 2$ to move to $k = 1$ amounts to

$$I_2 = r_2 x_2 + y_2 - (r_1 x_2 + y_2 - t).$$

Since by assumption $x_1 = X_1^0 = X_2^0 = x_2$, it follows that the average incentive to move $(I_1 + I_2)/2$ is equal to t and, hence, positive. Therefore, for any pair r_1 and r_2 of land rents at the two locations, at least one consumer would have an incentive to move. An equilibrium configuration does not exist.

4. FRACTIONAL ASSIGNMENT

Section 3 leads to the conclusion that a general result on the existence of equilibrium for models with location choice can be hoped for only if the number of consumers is large enough to make fractional assignments a meaningful approximation. Indeed, existence has been established in this case for various models of location choice (see [49], [29, part III], [30], [31] and [34]). Their approach based on fixed point theorems is not reproduced in this survey. We do, however, prove existence for the case of separable utility functions in Section 4.2 by solving an associated maximum problem. Recall that, by definition, the assignment problem consists of maximizing export at a fixed supply z of public goods. Since the whole of Section 4 deals with this problem, to simplify notation, z will not be written as an argument.

4.1 The second welfare theorem

The first welfare theorem (Theorem 1) gives a sufficient condition in terms of sustainability for an allocation to solve the fractional assignment problem. The condition turns out to be necessary as well provided, of course, that utility functions are quasi-concave in private consumption. This is the second welfare theorem for our model of residential choice.

THEOREM 2 *Given any solution* (n^A, x^A, y^A, S^A) *of the fractional assignment problem, there exist vectors* $r = (r_1, \ldots, r_K) \geq 0$ *and* $R = (R_1, \ldots, R_I)$ *of land rents and payments such that* (2.5.) *and* (2.6.) *hold.*

Sustaining land rents and payments are, of course, obtained as dual variables arising from the constraints of the fractional assignment problem. For differentiable utility functions, the proof can be based on the Kuhn–Tucker theorem (see [14]) whereas, for a more general proof based on the separating hyperplane theorem, the reader is referred to [47]. No proof will be given in this survey.

4.2. Equilibrium: the separable case

While sustainability of solutions of the assignment problem has been quite generally restored for fractional assignments (Theorem 2), the existence of Walrasian equilibria still remains to be established. This survey deals with existence for only the case of separable utility functions which can be written as $U_{ik}(x_{ik}, y_{ik}) = V_{ik}(x_{ik}) + y_{ik}$. Samuelson's [41] partial equilibrium approach is then adopted to our model of residential choice in the following way. For the *associated maximum problem,* the net social payoff here defined as

$$\sum_i \sum_k n_{ik}[V_{ik}(x_{ik}) + y^0_{ik}]$$

has to be maximized over all assignments $n = (\ldots, n_{ik}, \ldots)$ and vectors $x = (\ldots, x_{ik}, \ldots)$ of land consumption which satisfy the population and land constraints (2.1.) and (2.3.). Endowments enter the payoff's definition in order to net out commuting costs. Since the range of the associated maximum problem fails to be compact—all fractional assignments qualify—the existence of a solution should not be taken for granted but can be established under the proper set of assumptions. This technical problem will not be discussed further. We rather assume that a solution of the maximum problem, denoted by (n^M, x^M) does exist. Let π_i and r_k denote the variables dual to the population and land constraints (2.1.) and (2.3.), respectively. For differentiable utility functions, the first order conditions sustaining the solution of the associated maximum problem are computed in the familiar way as

$$n_{ik}^M[V'_{ik}(x_{ik}^M) - r_k] \leq 0 \qquad (4.1.)$$

with equality prevailing if $x_{ik}^M > 0$ and

$$V_{ik}(x_{ik}^M) + y_{ik}^0 - r_k x_{ik}^M - \pi_i \leq 0 \qquad (4.2.)$$

with equality prevailing if $n_{ik}^M > 0$.

Such a solution of the maximum problem can be completed to a Walrasian equilibrium of our model of location choice (closed economy, no public spending) provided that the level y_{ik}^M of consumption of the composite good as calculated from the budget equation

$$r_k x_{ik}^M + y_{ik}^M = y_{ik}^0 + \sum_l r_l x_{il}^0$$

is nonnegative (see Section 1) for all types i present at $k(n_{ik}^M > 0)$. In this case, demand of a utility maximizing consumer of type i living at location k amounts to (x_{ik}^M, y_{ik}^M) as follows from (4.1.). Moreover, consumers have no incentive to move because, as a consequence of (4.2.), they obtain the same utility level at all locations where their type is present. The solution of the associated maximum problem extended in the above way leads to a Walrasian equilibrium and, hence, existence is established.

To practise the method, let us reconsider the example of non-existence (first specification) as discussed in Section 3. Suppose the consumer's surplus from moving out (area ABCD of Figure 1) exceeds commuting costs t. No integer assignment can then be sustained in the sense of Theorem 1 and, hence, an integer equilibrium does not exist. To find the fractional equilibrium let x_1^M and x_2^M be the levels of land consumption emerging as solution of the associated maximum problem. The corresponding shadow prices for land at the two locations are denoted by r_1 and r_2. For the solution of the maximum problem, the surplus from moving out (area abcd of Figure 1) must be equal to commuting costs t as follows from first order conditions (4.1.) and (4.2.). This means that, under the configuration of Figure 1, less than one consumer live at the inner and more than one at the outer location. Only such an assignment would be consistent with equilibrium. To make such fractional assignments meaningful, a large number of consumers has to be assumed. The conclusion could then by restated as less than half of the population lives, in equilibrium, at the inner location.

4.3. The club assignment problem

In this section, the assignment problem is discussed from the "within club" view with still a fixed supply of public goods. The system $k = 1, 2, \ldots, K$ of locations, here referred to as an *agglomeration*, must be seen as part of a larger area such as the whole country. From the agglomeration's view, no binding population constraint exists because there is free emigration and immigration from and to the rest of the country. From the club viewpoint, the objective is to maximize utility per head. For the case of only one consumer type, the meaning of this requirement is obvious but not if there exist different types. Arnott and Stiglitz [3] postulate a club welfare function defined on the utility levels per head for all types. In this survey, an approach more in line with Paretian welfare economics will be taken. Following [46], an allocation is called *club efficient* if the per head utility allocation cannot be improved upon for all types at the same time by a possibly different composition of population. Put differently, imagine a utility frontier (in levels per head) plotted for every population structure. Club efficient allocations will then appear on the outer envelope of all these frontiers.

In this section, again, we deal with the more general case of open economies (see Section 2.1). A *club allocation* then consists of an assignment $n^C = (\ldots, n_{ik}^C, \ldots)$, an array of consumption vectors $x^C = (\ldots, x_{ik}^C \ldots)$ and $y^C = (\ldots, y_{ik}^C, \ldots)$ and a level of exports

S^C, such that (2.2.), (2.3.) and (2.4.) hold. A club allocation differs from our earlier notion of allocation in that the population constraint (2.1.) is no longer considered to be binding. The *club assignment problem* consists of maximizing the level S^C of exports over all club allocations consistent with the given utility allocation u. The notion of efficiency arising with solutions of the club assignment problem is, again, if anything more general and notationally simpler to handle than the concept of club efficiency (see Section 2.1). In particular, if preferences are monotonic in the composite good, any club efficient allocation has to solve the corresponding club assignment problem. The following theorem contains both a first and a second welfare theorem for the fractional club assignment problem. Solutions of the problem can be fully characterized in terms of sustainability by a system of land rents.

THEOREM 3 *A club allocation* (n^C, x^C, y^C, S^C) *solves the fractional club assignment problem if and only if there exists a vector* $r = (r_1, \ldots, r_K) \geq 0$ *of land rents such that*

$$S^C + C = \sum_k r_k X_k^0 \tag{4.3.}$$

and, for any bundle with $U_{ik}(x_{ik}, y_{ik}) \geq u_i$,

$$r_k x_{ik} + y_{ik} \geq y_{ik}^0. \tag{4.4.}$$

Recall the conditions of sustainability (2.5.) and (2.6.) appearing in Theorems 1 and 2. For the particular case of zero payments $(R = 0)$, these conditions are identical to (4.3.) and (4.4.) of the above theorem. Taking this into account, Theorem 3 becomes rather obvious. By dropping the population constraint (2.1) from the (total economy) assignment problem, the corresponding multiplier has to disappear. For a proof based on the separating hyperplane theorem, the reader is referred to [47].

For the remaining part of this section we concentrate on the club assignment problem for closed agglomerations (i.e. $S^C = 0$). In this case, for an allocation to solve the club assignment problem, aggregate land rent $\Sigma_r r_k X_k^0$ must just cover public expenditures (see (4.3.)). This result is a general version of the *Henry George Theorem* (see introduction) and allows to study the set of utility allocations consistent with club efficiency. The following two examples may serve as an illustration.

Suppose, first, the agglomeration consists of only one location (i.e. $K = 1$). The Henry George Theorem then implies that the land rent must be equal to $r = C/X^0$. Moreover, according to (4.4.), the income to be spent by consumers of type i is, at a solution of the club assignment problem, equal to their endowment y_i^0 of the composite good. Since prices (rent) and incomes are given, the utility allocation u^C consistent with club efficiency as well as the per capita land consumption for every consumer type are uniquely determined. It then follows from Theorem 3 that, for u^C, the club assignment problem can be solved at one-type as well as at fully mixed assignments (all types present) whereas, for utility allocations $u \geq u^C$, type i may be present ($n_i > 0$) only if $u_i = u_i^C$. We point out that the utility allocation u^C depends, of course, on the given supply z of public goods at which the club assignment problem is to be solved.

Now, let us turn to the case of agglomerations which are still closed in terms of the numeraire good ($S^C = 0$) but which are composed of two locations ($K = 2$). The Henry George Theorem (4.3.) requires public expenditures to be covered from land rent. Let α_k denote the fraction of revenues raised at location k hence $\alpha_1 + \alpha_2 = 1$. Land rent at k must then be $r_k = \alpha_k\ C/X_k^0$. Not every partition (α_1, α_2) of costs is consistent with club efficiency. Small values of α_1 lead to low rents at $k = 1$, such that nobody chooses to live at $k = 2$, where rents are high. For values of α_1 close to one, nobody wants to live at $k = 1$. Only values in between allow for solutions of the club assignment problem as we now want to show for the case of two consumer types ($I = 2$).

Starting from low values of α_1 inconsistent with club efficiency, α_1 is gradually increased to the point where the first type, say $i = 2$, feels indifferent between the two locations. At the corresponding utility allocation $u^I = (u_1^I, u_2^I)$, the club assignment problem can be solved with consumers of type $i = 2$ only or with a mixed agglomeration provided that consumers of type $i = 1$ reside at $k = 1$ only. It follows that u_2^I is the highest utility level per head attainable if consumers of type $i = 2$ live on their own, whereas u_1^I is the absolutely highest level for type $i = 1$ consistent with club efficiency. To realize u_1^I, however, consumers of type 2 have to fully occupy location $k = 2$ at a relatively low level of utility. Increasing fraction α_1 further and assigning type $i = 1$ to location $k = 1$ and type $i = 2$ to location $k = 2$ only (mixed agglomeration) still leads to solutions of the club assignment problem though, of course, at different utility levels. Fraction α_1, however, cannot be increased beyond the point where consumers of type $i = 1$ feel indifferent between the two locations. At the corresponding utility allocation $u^{II} = (u_1^{II}, u_2^{II})$, the level u_1^{II} is the highest attainable by consumers of type $i = 1$ on their own, whereas u_2^{II} is the highest utility for type $i = 2$ consistent with club efficiency. To attain u_2^{II}, consumers of type $i = 1$ have to be present at relatively low utility, fully occupying location $k = 1$.

The extreme values u^I and u^{II} of utility allocations depend, again, on the supply z of public goods at which the club assignment is to be solved. At a fixed supply, say z^0, the utility allocation moves from u^1 to u^{II} (see Figure 2) as the fraction α_1 of costs borne by location $k = 1$ increases. Curve I I of Figure 2 depicts the utility allocation u^I as resulting from different levels z of public supply whereas curve II II does the same for the utility allocation $u^{II.}$ Utility allocations on

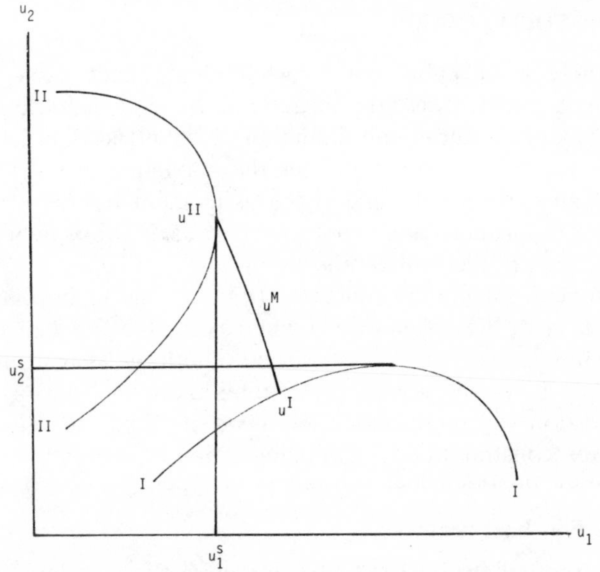

FIGURE 2 Utility allocations consistent with club efficiency ($K = 2$, $I = 2$).

II II can be realized by consumers of type 1 on their own at the proper supply of public goods. Similarly, consumers of type 2 can obtain utility allocations on I I without needing consumers of the other type to share costs. Let z_1^s and z_2^s denote the supply of public goods which leads to the highest utility levels u_1^s and u_2^s, respectively, attainable in agglomerations with only one consumer type. In Figure 2, obviously, parameters are chosen such that $z_1^s = z^0$ and such that $z_2^s \neq z_1^s$ because $u_1^s = u_1^{II}$ and $u_2^I < u_2^s$. For an agglomeration with type $i = 2$ only being present, the level of supply z_1^s would not be optimal.

The example of Figure 2 will be further discussed in Section 5.2 which deals with the notion and existence of Tiebout equilibria. The case with more than two types and locations could be analysed along similar lines though, of course, at greater combinatorial effort. In any case, our findings on the assignment problem at a fixed supply of public goods turn out to be a useful first step towards understanding the more complicated first best problem to be studied in the next chapter.

5. LOCAL PUBLIC GOODS

The supply z of public goods while being kept fixed for the assignment problem is to be determined endogenously in the present chapter. Recall the definition given in Section 2.1 of the function $s(u, n, z)$ which expresses the maximum level of exports over all allocations with underlying assignment $n = (\ldots, n_{ik}, \ldots)$ of people to locations and supply $z = (\ldots, z_k, \ldots)$ of public goods consistent with the utility allocation $u = (\ldots, u_i, \ldots)$. For the assignment problem, the function $s(u, n, z)$ has to be maximized with respect to the assignment n whereas for the *first best problem* the function is maximized with respect to both the assignment n and the supply of public goods z. A "total economy" and a "within club" version will be distinguished again according to whether the population constraint (2.1.) is binding or not.

5.1. The first best problem

As an intermediate step, we first deal with the so-called *Lindahl-Samuelson problem* which consists of maximizing the level of exports $s(u, n, z)$ with respect to the supply z of public goods at a fixed assignment n. Since there is no location choice this problem is essentially aspatial in nature. In order to solve this problem we suppose that utility functions are quasi-concave in private and public consumption and that costs $C(z)$ of producing public goods are convex as a function of z. Solutions of the Lindahl-Samuelson problem can then be sustained by land rents, personalized (Lindahl) prices for public goods and a system of payments as stated in Theorem 4. In other words, a first and second welfare theorem for the Lindahl-Samuelson problem holds.

THEOREM 4 *The allocation* $(n^L, x^L, y^L, z^L, S^L)$ *solves the Lindahl-Samuelson problem if and only if there exists a vector* $r = (r_1, \ldots, r_K) \geq 0$ *of land rents and, for all locations* k *where consumer type* i *is present* $(n_{ik}^L > 0)$, *a personalized Lindahl price vector* $q_{ik} \geq 0$ *and a level* R_{ik} *of payments such that, for* $Q = \sum n_{ik}^L q_{ik}$,

$$S^L + \sum_i \sum_k n_{ik}^L R_{ik} = \sum_k r_k X_k^0 + Q z^0 - C(z^0) \qquad (5.1.)$$

$$Qz - C(z) \leq Qz^0 - C(z^0) \quad \textit{for all} \quad z \geq 0 \qquad (5.2.)$$

and, for any personalized bundle with $U_{ik}(x_{ik}, y_{ik}, z_{ik}) \geq u_i$,

$$r_k x_{ik} + y_{ik} + q_{ik} z_{ik} \geq y_{ik}^0 + R_{ik}. \qquad (5.3.)$$

Theorem 4 is a general version of Samuelson's [42] condition that the sum of marginal rates of substitution over all consumers should, at the optimum supply of public goods, be equal to the marginal rate of transformation. Condition (5.2.) requires that the Lindahl firm producing public goods behaves as a price taking profit maximizer, whereas (5.3.) implies that q_{ik} is the marginal rate of substitution between public goods and private consumption of the composite good for consumers of type i living at k. On the right-hand side of (5.1.) profits of the Lindahl firm appear as a source of income in addition to aggregate land rent. For a general proof of this rather classical result the reader is referred to Milleron [35].

For a given utility allocation u, let us now consider an allocation $(n^*, x^*, y^*, z^*, S^*)$ which solves the first best problem. Such an allocation, of course, will also be a solution of both the assignment and the Lindahl-Samuelson problem. If a function attains its maximum with respect to two sets of variables, it has to attain there its maximum with respect to each of them separately. It then follows from Theorems 2 and 4 that the solution of the first best problem must be sustainable in the sense of both theorems. Assume, for simplicity, that the sustaining vectors of land rents appearing in the two theorems are the same as would certainly be the case if utility and cost functions are differentiable. Conditions (2.6.) and (5.3.) then imply that the payments appearing in Theorems 2 and 4 are related in the following way

$$R_{ik} = R_i + q_{ik} z^*. \tag{5.4.}$$

Suppose for a moment that consumers of type i living at k are required to pay a Lindahl tax amounting to $q_{ik} z^*$ for public consumption. The system R_{ik} of payments must then compensate for differences in tax levels over locations as follows from (5.4.). Otherwise, efficiency losses would arise from distortions in location choice. Such Lindahl taxes, however, need not be raised because, according to (2.5.), aggregate land rent if taxed away would be sufficient to cover public expenditures C, the level of exports S^L and the sum $\sum N_i^0 R_i$ needed for payments. In particular, for a closed economy ($S^L = 0$), aggregate rent net of public expenditures corresponds to net income from land ownership.

To summarize, solutions of the first best problem can be sustained by a system of prices, rents and payments such that conditions (2.5.), (2.6.), (5.1.), (5.3.) and (5.4.) hold. Sustainability is necessary for an allocation to solve the first best problem which, in this sense, is shown to allow for a second welfare theorem (see [14], [7] for similar results). Unfortunately, a corresponding first welfare theorem does not generally hold. The above conditions of sustainability are not sufficient for an allocation to be a solution. Of course, any allocation $(n^*, x^*, y^*, z^*, S^*)$ sustained in the above sense must solve both the assignment and the Lindahl-Samuelson problem as follows from Theorems 1 and 4. In other words, it must be true that $S^* = s(u, n^*, z^*) \geq s(u, n, z^*)$ for all assignments n as well as $S^* = s(u, n^*, z^*) \geq s(u, n^*, z)$ for any other supply z of public goods. But since $s(u, n, z)$ is not, in general, concave as a function of n and z together, a different assignment and level of public spending could still exist which would lead to a higher level of exports than S^*. Allocations which are not first best may be sustainable.

The occurrence of both a variable supply of public goods and of location choice leads to nonconvexities beyond what can be repaired by resorting to fractional instead of integer assignments. Formally speaking, while $s(u, n, z)$ is concave with respect to n as well as to z, it need not be concave with respect to both of them as can be shown by examples easy to construct (see [55] for a related discussion). Whether the problem should simply be dismissed as a mathematical curiosity remains to be seen. In any case, one must be aware of such a possibility particularly if the inefficiency of spatial equilibrium is claimed. The "within club" view could, of course, be discussed similarly. While allocations solving the first best club problems are sustainable at zero payments ($R_i = 0$), sustainability does not seem sufficient for a club allocation to be a solution.

5.2. Tiebout equilibrium

Tiebout [61], in a seminal paper, has argued that no free rider problem arises with the supply of local public goods whose incidence of utility, by definition, is spatially limited. Preferences are revealed by residential choice ("Voting by Feet") and, as a consequence, resources for public use will be spent efficiently (see [64] for further discussion of the idea). Tiebout's intriguing idea has led to many controversies on the concept, efficiency and existence of equilibrium. For a recent and comprehensive critique of the idea, the reader is referred to Bewley [8] and the references given there.

Bewley's rigorous notion of a Tiebout equilibrium can be adopted to our framework of location choice in the following way. Local governments or what Arnott [2] calls land developers are assumed to earn rents from organizing local communities here referred to as agglomerations. Consumers choose freely where to live and, hence, give rise to competition among developers for inhabitants. Developers are numerous enough to perceive the supply of consumers at the utility allocation obtained outside their domain as infinitely elastic. They behave as *utility takers* (see Section 2.2), maximizing the rental income by regulating the agglomeration's population structure and by choosing the supply of public goods properly. If the level of exports is given the meaning of rental income then utility taking developers can be visualized as solving the club version of the first best problem. With free entry of developers as Bewley seems to assume, rental incomes are driven down to the zero level. It then follows that only utility allocations which allow to solve the club version of the first best problem at zero exports would qualify for a Tiebout equilibrium provided, of course, that the given national population can be divided into agglomerations of optimum structure and size. This, however, may not be possible in which case equilibrium fails to exist.

For illustration, let us reconsider the examples discussed in Section 4.3. If agglomerations, to begin with, consist of only one location ($K = 1$), the maximum utility level per head for every type can be realized, at the proper size and supply of public goods, within one-type agglomerations. If the given national population can be divided into such agglomerations, then not only does a Tiebout equilibrium exist, but consumers of different types come to live in different agglomerations. If agglomerations consist of two locations ($K = 2$) and, for simplciity, suppose that there exist only two consumer types ($I = 2$) then the maximum utility level per head for a given type consistent with club efficiency can only be realized within mixed agglomerations. Put differently, the highest level of utility u_1^S and u_2^S which consumers of each type can obtain by themselves will be dominated by utility allocations such as u^M in Figure 2 which have to be realized within mixed agglomerations. If a Tiebout equilibrium exists, then consumers will live in mixed communities (see Berglas [6] and Berglas and Pines [7] for similar findings).

The integer problem with respect to the number of agglomerations, however, may still arise in which case no equilibrium in Bewley's sense will exist. Existence becomes more likely if rental incomes of developers are allowed to be positive (see [47]). And, as a matter of fact, Bewley's assumption seems questionable. Recall the quite similar situation of Cournot oligopoly with free entry where entry, in general, is predicted to stop at positive levels of profit. The number of incumbents is determined such that earnings would drop below zero if one more were to enter. In this sense, positive incomes of developers would not be inconsistent with equilibrium. In particular, if the number of potential sites for agglomerations is limited or if there exists diversity of the resource base, positive incomes are rather to be expected. Such incomes cause problems for the notion of Tiebout equilibrium unless they can be assumed as being absorbed by absentee landlords. In general, however, some central authority is needed which redistributes these rental incomes among consumers in the form of lump-sum payments (see [24]). While every efficient allocation can be sustained in this way as follows from the results presented in the survey, a concept of equilibrium is only to be defined with respect to a given scheme of payments. The problem of assigning people to locations, however, can still be solved by utility taking developers or organizers as has been shown in Corollary 1.2.

6. CONCLUDING REMARKS

In traditional models without location choice, general equilibrium and normative analysis are closely related subjects. The relation is given by the full duality arising from the two welfare theorems. Every equilibrium allocation has to be efficient (first welfare theorem) and every efficient allocation can be sustained as an equilibrium with the proper redistribution of initial endowments (second welfare theorem). To be sure, competitive equilibrium should primarily be seen as a concept of positive theory. But the normative approach leads to sustaining price systems in an equally natural way. Here, prices simply play the role of dual variables.

With location choice, however, welfare duality will at least partly break down. The degree to which this happens has been explored in the present survey. It is shown that, at a fixed supply of public goods, full duality can be restored by assuming a continuum of agents. This assumption seems to be a plausible approximation for a model of residential choice where mobile agents are all relatively small. The assumption seems less plausible if location choosing agents are plants producing output under economies of scale. Moreover, even with a continuum of agents, there will be no full duality if the supply of public goods is to be determined endogenously. More precisely, while the assignment problem at a fixed supply of public goods as well as the Lindahl-Samuelson problem at a fixed assignment both allow for full welfare duality, this need no longer be the case if assignment and public expenditures have to be determined simultaneously. Even more serious difficulties will arise if some general form of externality has to be taken into account.

Without full welfare duality, normative theory becomes a topic of little elegance. Difficulties arise not only from characterizing optimum solutions, but also from establishing direct links between normative and positive theory. If the optimum cannot be sustained as an equilibrium then the question will be which allocation mechanism leads to the optimum. In this case, a positive theory which simply explains reality as if it were heading for some optimum does not seem very convincing. Nevertheless, it might be of use to know something about the optimum even though no results of great generality can be expected to hold under less than full welfare duality. I have done a few tentative steps in this direction (see [48], and I strongly believe that further research would be highly desirable in the area of normative theory. The optimum might be difficult to characterize and to reach. But it can serve as a distinctive point of reference. A truly positive theory, on the other hand, should predict allocations as the outcome of explicit interaction between households, firms and governmental authorities. To begin with, one should probably look for some simple *ad hoc* rules of behavior. Theoretical analysis will then lead to predictions based on such rules. As long as behavioral assumptions cannot be tested empirically, positive theories will survive or be refuted according to their predictive power.

The present survey discusses the limits of general equilibrium theory for the analysis of spatial economies with location choice. To proceed beyond these limits, normative as well as positive approaches should be tried despite the fact that theories of sweeping generality cannot be expected to exist. One should rather be prepared to encounter a situation similar to that known from the theory of industry structure. While normative analysis turns out be painful if scale economies are present and if entry occurs (see [4]), the positive approach to oligopoly has led to an equally confusing variety of models and results. This seems the price to be paid for dealing with situations which do not allow for complete welfare duality and which, therefore, are beyond the reach of general equilibrium theory.

References

[1] Arnott, R., "Optimal City Size in a Spatial Economy", *Journal of Urban Economics,* **6** (1979), 65–89.
[2] Arnott, R. J., "Urban Agglomeration and its Efficiency: A Critical Review of the Literature with Some Extensions", Discussion Paper, Department of Economics, Queen's University, January 1983.
[3] Arnott, R. J. and J. E. Stiglitz, "Aggregate Land Rents, Expenditure on Public Goods, and Optimal City Size", *Quarterly Journal of Economics,* **93** (1979), 471–500.
[4] Baumol, W. J., J. C. Panzar and R. D. Willig, *Contestable Markets and the Theory of Industry Structure.* New York: Harcourt Brace Jovanovich Inc., 1982.
[5] Beckmann, M. J., *Location Theory.* New York: Random House, 1968.
[6] Berglas, E., "Distribution of Tastes and Skills and the Provision of Local Public Goods", *Journal of Public Economics,* **93** (1976), 406–423.
[7] Berglas, E. and D. Pines: "Clubs, Local Public Goods and Transportation Models: A Synthesis", *Journal of Public Economics,* **15** (1981), 141–162.
[8] Bewley, T. F.: "A Critique of Tiebout's Theory of Local Public Expenditures", *Econometrica,* **49** (1981), 713–740.
[9] Buchanan, J. M.: "An Economic Theory of Clubs", *Economica,* **32** (1965), 1–14.
[10] Debreu, G.: *Theory of Value.* Cowles-Foundation Monograph 17, 1959.
[11] Debreu, G. and H. Scarf: "A Limit Theorem on the Core of an Economy", *International Economic Review,* **4** (1963), 235–246.
[12] Dixit, A., "The Optimum Factory Town", *Bell Journal of Economics,* **4** (1973), 637–654.
[13] Enke, S., "Equilibrium among Spatially Separated Markets: Solution by Electric Analogue", *Econometrica,* **19** (1951), 40–47.
[14] Flatters, F., V. Henderson and P. Mieszkowski: "Public Goods, Efficiency, and Regional Fiscal Equalization", *Journal of Public Economics,* **3** (1974), 99–112.
[15] Greenberg, J., "Local Public Goods with Mobility: Existence and Optimality of a General Equilibrium", *Journal of Economic Theory,* **30** (1983), 17–33.
[16] Hamilton, B. W., "Indivisibilities and Interplant Transportation Cost: Do They Cause Market Breakdown?", *Journal of Urban Economics,* **7** (1980), 31–41.
[17] Hartwick, J. M., "The Pricing of Goods and Agricultural Land in Multiregional General Equilibrium", *Southern Economic Journal,* **39** (1972), 31–45.
[18] Hartwick, J. M., "The Location of Firms and General Spatial Price Equilibrium", *Weltwirtschaftliches Archiv,* **108** (1972), 462–482.
[19] Hartwick, J. M., "Price Sustainability of Location Assignments", *Journal of Urban Economics,* **1** (1974), 147–160.
[20] Hartwick, J. M., "The Henry George Rule, Optimal Population, and Interregional Equity", *Canadian Journal of Economics,* **13** (1980), 695–699.
[21] Heffley, D. R., "The Quadratic Assignment Problem: A Note", *Econometrica,* **40** (1972), 1155–1163.
[22] Heffley, D. R., "Efficient Spatial Allocation in the Quadratic Assignment Problem", *Journal of Urban Economics,* **3** (1976), 309–322.
[23] Heffley, D. R., "Competitive Equilibria and the Core of a Spatial Economy", *Journal of Regional Science,* **22** (1982), 423–440.
[24] Helpman, E., "On Optimal Community Formation", *Economic Letters,* **1** (1978), 289–293.
[25] Helpman, E. and A. L. Hillman: "Two Remarks on Optimal Club Size", *Economica,* **44** (1977), 293–296.

[26] Herbert, J. D. and B. H. Stevens: "A Model for the Distribution of Residential Activity in Urban Areas", *Journal of Regional Science,* **2** (1960), 21–36.

[27] Kanemoto, Y., *Theories of Urban Externalities.* Amsterdam: North-Holland Publishing Company, 1980.

[28] Kanemoto, Y., "Externalities in Space", *Fundamentals of Pure and Applied Economics,* forthcoming.

[29] Karmann, A. J., *Competive Equilibria in Spatial Economies.* Königstein/Ts.: Anton Hain, 1981.

[30] Karmann, A. J., "Spatial Barter Economies Under Locational Choice", *Journal of Mathematical Economics,* **9** (1982), 259–274.

[31] Karmann, A. J., "Space-Time Economies Under Free Mobility", *Regional Science and Urban Economics,* **14** (1984), 303–316.

[32] Kennedy, M., "An Economic Model of the World Oil Market", *The Bell Journal of Economics and Management Science,* **5** (1974), 540–577.

[33] Koopmans, T. C. and M. J. Beckmann, "Assignment Problems and the Location of Economic Activities", *Econometrica,* **25** (1957), 53–76.

[34] Mazzoleni, P. and Montesano, A., "General Competitive Equilibrium of the Spatial Economy", *Regional Science and Urban Economics,* **14** (1984), 285–302.

[35] Milleron, J. C., "Theory of Value with Public Goods", *Journal of Economic Theory,* **5** (1972), 419–477.

[36] Mills, E. S. and J. Mackinnon: "Notes on the New Urban Economics", *Bell Journal of Economics,* **4** (1973), 593–601.

[37] Miron, J. R. and P. Sharke: "Non-Price Information and Price Sustainability in the Koopmans-Beckmann Problem", *Journal of Regional Science,* **21** (1981), 117–122.

[38] Mirrlees, J. A., "The Optimum Town", *Swedish Journal of Economics,* **74** (1972), 114–135.

[39] Papageorgiou, Y. Y., "Agglomeration", *Regional Science and Urban Economics,* **9** (1979), 41–59.

[40] Richter, D. K., "Weakly Democratic Regular Tax Equilibria in a Local Public Goods Economy with Perfect Consumer Mobility", *Journal of Economic Theory,* **27** (1982), 137–162.

[41] Samuelson, P. A., "Spatial Price Equilibrium and Linear Programming", *American Economic Review,* **42** (1952), 283–303.

[42] Samuelson, P. A., "The Pure Theory of Public Expenditure", *Review of Economics and Statistics,* **36** (1954), 387–389.

[43] Sandler, T. and J. T. Tschirhart, "The Economic Theory of Clubs: An Evaluative Survey", *Journal of Economic Literature,* **18** (1980), 1481–1521.

[44] Schotter, A., "On Urban Residential Stability and the Core", *Regional Science and Urban Economics,* **7** (1977), 321–338.

[45] Schweizer, U., "Edgeworth and the Henry George Theorem: How to Finance Local Public Projects", in: Thisse, J.-F. and H. G. Zoller (eds.) *Locational Analysis of Public Facilities.* Amsterdam: North-Holland, 1983.

[46] Schweizer, U.: "Efficient Exchange with a Variable Number of Consumers", *Econometrica,* **51** (1983), 575–584.

[47] Schweizer, U., "Fiscal Decentralization under Free Mobility", *Regional Science and Urban Economics,* **14** (1984), 317–330.

[48] Schweizer, U., "Theory of City System Structure", *Regional Science and Urban Economics,* **15** (1985), 159–180.

[49] Schweizer, U., P. Varaiya and J. Hartwick: "General Equilibrium and Location Theory", *Journal of Urban Economics,* **3** (1976), 285–303.

[50] Serck-Hanssen, J., "The Optimal Number of Factories in a Spatial Market", in: Bos, H. (ed.) *Toward Balanced International Growth.* Amsterdam: North-Holland Publishing Company, 1969.
[51] Shapley, L. S. and M. Shubik: "The Assignment Game I: The Core", *International Journal of Game Theory,* **1** (1972), 111–130.
[52] Smith, V. L., "Minimization of Economic Rent in Spatial Price Equilibrium", *Review of Economic Studies,* **30** (1963), 24–31.
[53] Starrett, D., "Principles of Optimal Location in a Large Homogenous Area", *Journal of Economic Theory,* **9** (1974), 418–448.
[54] Starrett, D., "Market Allocations of Location Choice in a Model with Free Mobility", *Journal of Economic Theory,* **17** (1978), 21–37.
[55] Stiglitz, J. E., "The Theory of Local Public Goods", in: Feldstein, M. and R. Inman (eds.): *The Economics of Public Services.* London: Macmillan, 1977.
[56] Takayama, T. and G. G. Judge, "Equilibrium among Spatially Separated Markets: A Reformulation", *Econometrica,* **32** (1964), 510–524.
[57] Takayama, T. and G. G. Judge: "*Spatial and Temporal Price and Allocation Models*". Amsterdam: North-Holland, 1971.
[58] Takayama, T. and N. Uri: "A Note on Spatial and Temporal Price and Allocation Modeling: Quadratic Programming or Linear Complementarity Programming?", *Regional Science and Urban Economics,* **13** (1983), 455–470.
[59] Thore, S., "The Takayama-Judge Spatial Equilibrium Model with Endogenous Income", *Regional Science and Urban Economics,* **12** (1982), 351–364.
[60] Thore, S., "The Takayama-Judge Model of Spatial Equilibrium extended to Convex Production Sets", *Journal of Regional Science,* **22** (1982), 203–212.
[61] Tiebout, C., "A Pure Theory of Local Expenditures", *Journal of Political Economy,* **64** (1956), 416–424.
[62] Vickrey, W., "The City as a Firm", in: Feldstein, M. and R. Inman (eds.): *The Economics of Public Services.* London: Macmillan, 1977.
[63] Wildasin, D. E., "The Welfare Effects of Intergovernmental Grants in an Economy with Independent Jurisdictions", *Journal of Urban Economics,* **13** (1983), 147–164.
[64] Wildasin, D. E., "Urban Public Finance", *Fundamentals of Pure and Applied Economics,* in press.
[65] Wildasin, D. E., "Spatial Variation of the Marginal Utility of Income and Unequal Treatment of Equals", *Journal of Urban Economics,* to be published.

INDEX

FUNDAMENTALS OF PURE AND APPLIED ECONOMICS

Volume 1 (International Trade Section)
GAME THEORY IN INTERNATIONAL ECONOMICS
by John McMillan

Volume 2 (Marxian Economics Section)
MONEY, ACCUMULATION AND CRISIS
by Duncan K. Foley

Volume 3 (Theory of the Firm and Industrial Organization Section)
DYNAMIC MODELS OF OLIGOPOLY
by Drew Fudenberg and Jean Tirole

Volume 4 (Marxian Economics Section)
VALUE, EXPLOITATION AND CLASS
by John E. Roemer

Volume 5 (Regional and Urban Economics Section)
LOCATION THEORY
by Jean Jaskold Gabszewicz and Jacques-François Thisse, Masahisa Fujita, and Urs Schweizer

Volume 6 (Political Science and Economics Section)
MODELS OF IMPERFECT INFORMATION IN POLITICS
by Randall L. Calvert

Volume 7 (Marxian Economics Section)
CAPITALIST IMPERIALISM, CRISIS AND THE STATE
by John Willoughby

Volume 8 (Marxian Economics Section)
MARXISM AND "REALLY EXISTING SOCIALISM"
by Alec Nove

Volume 9 (Economic Systems Section)
THE NONPROFIT ENTERPRISE IN MARKET ECONOMICS
by Estelle James and Susan Rose-Ackerman

Volume 10 (Regional and Urban Economics Section)
URBAN PUBLIC FINANCE
by David E. Wildasin

Volume 11 (Regional and Urban Economics Section)
URBAN DYNAMICS AND URBAN EXTERNALITIES
by Takahiro Miyao and Yoshitsugu Kanemoto

Volume 12 (Marxian Economics Section)
DEVELOPMENT AND MODES OF PRODUCTION IN MARXIAN ECONOMICS: A CRITICAL EVALUATION
by Alan Richards

Volume 13 (Economics of Technological Change Section)
TECHNOLOGICAL CHANGE AND PRODUCTIVITY GROWTH
by Albert N. Link

Additional volumes in preparation
ISSN: 0191-1708